Analyzing Chromosomes

Barbara Czepulkowski

Leukaemia Sciences Laboratories
The Rayne Institute, London, UK

BIOS

© BIOS Scientific Publishers Limited, 2001

First published 2001

A CIP catalogue record for this book is available from the British Library.

ISBN 1 85996 188 6

BIOS Scientific Publishers Ltd
9 Newtec Place, Magdalen Road, Oxford OX4 1RE, UK
Tel. +44 (0)1865 726286. Fax +44 (0)1865 246823
World Wide Web home page: http://www.bios.co.uk/

Published in the United States of America, its dependent territories and Canada by Springer-Verlag New York Inc., 175 Fifth Avenue, New York, NY 10010–7858, in association with BIOS Scientific Publishers Ltd

Production Editor: Andrea Bosher
Typeset by Creative Associates, Oxford, UK
Printed by The Cromwell Press, Trowbridge, UK

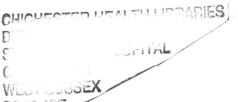
Analyzing
Chromosomes

Series Advisors:

Rob Beynon UMIST, Manchester, UK
Chris Howe Department of Biochemistry, University of Cambridge, Cambridge, UK

Monoclonal Antibodies
PCR
Separating Cells
Analyzing Chromosomes

Forthcoming titles

Biological Centrifugation
Gene Mapping
Reconstructing Molecular Evolutionary Trees

Contents

Chapter 3 Staining and banding of chromosome slides   49

Barbara Czepulkowski

Chapter 4 Types of abnormalities observed in chromosomes   69

Barbara Czepulkowski

Chapter 5 Approaching analysis   91

Barbara Czepulkowski

Chapter 6 Constitutional chromosome abnormalities   109

Barbara Czepulkowski and Karen Saunders

Chapter 7 Acquired chromosome abnormalities observed in malignancy   125

Barbara Czepulkowski

Chapter 8 New complementary techniques   159

Karen Saunders and David Jones

Appendix I   183

Appendix II   191

Index   199

The color section can be found between pages 18 and 19

 **The Basics**

Abbreviations

ALL	acute lymphoblastic leukemia
AML	acute myeloblastic leukemia
APML	acute promyelocytic leukemia
ATL/L	adult T cell leukemia/lymphoma
ATRA	all-*trans* retinoic acid
BrdU	bromodeoxyuridine
CGH	comparative genomic hybridization
CGL	chronic granulocytic leukemia
CLL	chronic lymphocytic leukemia
CML	chronic myeloid leukemia
CMMoL	chronic myelomonocytic leukemia
CNL	chronic neutrophilic leukemia
CNS	central nervous system
CVS	chorionic villus sample
DA	distamycin A
DAPI	4,6-diamino-2-phenyl-indole
DIC	disseminated intravascular coagulation
DLBCL	diffuse large B cell lymphoma
DNA	deoxyribonucleic acid
EBV	Epstein–Barr virus
FAB	French–American–British
FdU	fluorodeoxyuridine
FISH	fluorescent *in situ* hybridization
FL	follicular lymphoma
HCL	hairy cell leukemia
HCL-v	HCL variant
HD	Hodgkin's disease
Ig	immunoglobulin
jCML	juvenile CML
LGL	large granular cell lymphocytosis
MCL	mantle cell lymphoma
MDS	myelodysplastic syndrome
MPD	myeloproliferative disorder
mRNA	messenger RNA
MTX	methotrexate
NHL	non-Hodgkin's lymphoma
NOR	nuclear organizer region
PBS	phosphate-buffered saline
PCL	plasma cell leukemia

PCR	polymerase chain reaction
PHA	phytohemagglutinin
PLL	prolymphocytic leukemia
PMA	4-phorbol-12-myristate-13-acetate
POC	products of conception
PRINS	primed *in situ* hybridization
PRV	polycythemia rubra vera
PWM	pokeweed mitogen
REAL	Revised European–American Lymphoma classification
RNA	ribonucleic acid
rRNA	ribosomal RNA
SARs	scaffold attachment regions
SLVL	splenic lymphoma with villous lymphocytes
T-NHL	T cell non-Hodgkin's lymphoma
TCR	T cell receptor
TPA	12-0 tetradecanoylphorbol-13-acetate
UPD	uniparental disomy
yacs	yeast artificial chromosomes

Contributors

Karen Saunders, Leukaemia Science Laboratories, The Rayne Institute, London, UK

David Jones, Leukaemia Science Laboratories, The Rayne Institute, London, UK

Introduction

Barbara Czepulkowski

Throughout my employment as a clinical cytogeneticist, I have been constantly asked by fellow hospital workers, friends, family and other peripheral types to explain exactly what I do as a cytogeneticist. This explanation would never be easy! In fact even when one begins training a fledgling cytogeneticist, the difficulty in getting across the reasons for performing certain acts and the varying approaches to analysis can be frustrating. With clinicians and research workers I found that a small volume by Bishop and Cook, published by Heinemann Medical Books as long ago as 1966, proved extremely useful and, although faded, yellowed and bent, the volume, with its concise explanations, has proved invaluable as a teaching tool.

It had always been in the recesses of my mind that a new up-to-date version of this type of volume should be embarked upon. However, each time I thought this to myself, other events would take over and the 'mythical' volume would be returned to the shelf. In the not too distant past, I was asked by a journal to provide a concise page or two about analyzing chromosomes and, although it seemed an impossible task, I did my best. (It was the equivalent to trying to write the complete works of Shakespeare on a grain of rice!)

Almost at the very same moment I received a call from Dr David Rickwood asking me whether I would be interested in writing a volume (a simplistic approach to analyzing chromosomes) for would-be cytogeneticists and also research workers and clinicians. Naturally, I thought it was a brilliant idea. The rest, as they say, is history!

It is hoped that the ensuing pages will prove useful to all those who are not trained cytogeneticists, and even those who are, who like me may occasionally forget certain symptoms of varying syndromes. A simple quick reference guide to look up the relevant information would be a bonus.

Once I planned the volume I realized that it was going to prove more difficult than I thought to achieve the correct balance of information without becoming too technical for those not trained in cytogenetics. However, I hope that we have achieved this, as I have employed the brains and ideas of my colleagues at King's College Hospital. Special thanks are due to Karen Saunders, who not only reminded me of forgotten subjects in the prenatal and constitutional cytogenetics field, but kindly agreed to write the chapter on 'New complementary techniques' and made a major contribution to the methods and text in Chapter 2. I would also like to thank David Jones, who provided one of the sections on these new techniques. We hope that other centers, as well as King's College Hospital, will enjoy the contents as much as we enjoyed preparing them. I also give thanks to the many books and journals I have cited, as the information, which I have collated into this concise volume, proved invaluable.

This volume is not meant as a complete guide, nor as a substitute for all the excellent text books which are currently available that delve more deeply into the subject of cytogenetics. However, it is hoped that those without genetics degrees and intense training in cytogenetics may extract some useful and interesting information regarding the mystical world of chromosome analysis. The chapters take us through the introduction to the cell, preparing chromosomes from various tissues in clinical use, different banding techniques which enable us to distinguish the chromosomes and recognition of the patterns obtained. The next chapter offers a guide to analytical approaches and the problems one may encounter during analysis of chromosomes. The following two chapters discuss the different abnormalities associated with varying genetic syndromes in the constitutional field, and acquired changes which occur in malignant disease. The final chapter takes us through the colorful new molecular techniques that have enhanced our subject immensely.

It is difficult to choose a single acknowledgment which would justify all the people who have inspired me throughout my career. There are many who have 'kept my motor running' during traumas in life and kept the spirit free through troubled times, but many people (and creatures great and small) have contributed to the interesting journey. Because I do not want to omit anyone, I have dedicated this book to everyone that knows me and everyone that does not, because I hope they will come to know me during their journey through the pages of this book.

Appligene Oncor kindly sponsored the color pages, which were essential to demonstrate some of the new molecular techniques used in the cytogenetics field. I am indebted to their generous contribution, and offer my sincere thanks, as the relevant photographs would have been uninformative had they not been portrayed in color. I would also like to give many thanks to Angela Douglas of the Liverpool Women's Hospital for providing most of the photographs for Chapters 3 and 6 in a most efficient and speedy manner. Special thanks to Jonathan Ray and Victoria Oddie of BIOS Scientific Publishers for their perseverence and professionalism.

This book is especially for: Dudley (bin there, seen it, dun it); Mick (give me something to save) Blythe; Dave (still crawling from the wreckage) Edmunds; Suki (Devil); Kiri (Angel); and Revenge (simply the best). These proofs were intended reading for any 'rain-stopped-play' occurances at the Third Test at Headingley Cricket Ground on Saturday 19 August. We had tickets unused, as the magnificent performance by the England cricketers bought the test match to a conclusion in 2 days! Bravo!

Chapter 1

Introducing the cell

Barbara Czepulkowski

1.   Introduction

To understand the fundamentals involved when learning about chromosomes and their analysis, it is important to explore their origins, structure and functions. It is therefore prudent to begin at the beginning, to coin a phrase, and start with the 'home' of the chromosome, the cell. In all forms of life, some protoplasmic material exists, with a protective boundary surrounding it. Single-celled organisms possess only one such boundary, whereas more complex organisms contain many individual units or building blocks, composed of a number of cell types. However, the organization of all of these cells is similar in all organisms.

The cell usually consists of two distinct areas, the cytoplasm (which forms the major part of the protoplasmic material) and, lying within this area, a dark-staining body called the nucleus. In bacteria and blue-green algae, the nucleus is not separated from the cytoplasm by a discrete membrane, whereas in multicellular organisms the genetic material contained in the nucleus is separated from the cytoplasm by the nuclear membrane. In viral organisms, the only membrane present is the viral envelope enclosing all the viral genetic material.

2.   The cytoplasm

The cytoplasm contains a number of organelles, each undertaking various functions, and in eukaryotes (an organism which possesses a true nucleus) the structures named below can be observed under the electron microscope (see *Figure 1.1*).

- Mitochondria – small bodies with shelf-like internal layers where respiration and oxidation take place, providing cellular energy.
- Golgi apparatus – net-like staining bodies commonly found in secretory cells.
- Endoplasmic reticulum – cytoplasmic double-walled membrane, folded in layers, which appears to be connected to cell membranes.
- Ribosomes – associated with the endoplasmic reticulum, and playing a role in the synthesis of proteins.

Proteins are composed of one or more polypeptide molecules, which can be modified by various carbohydrate side chains or other chemical groups.

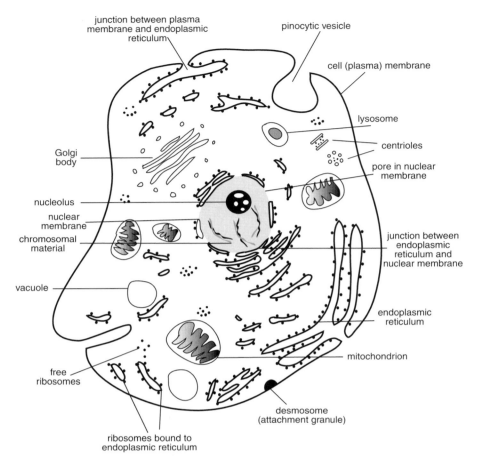

junction between plasma
membrane and endoplasmic
reticulum

pinocytic vesicle

cell (plasma) membrane

lysosome

centrioles

pore in nuclear
membrane

Golgi
body

nucleolus

nuclear
membrane

chromosomal
material

junction between
endoplasmic
reticulum and
nuclear membrane

vacuole

endoplasmic
reticulum

mitochondrion

free
ribosomes

desmosome
(attachment granule)

ribosomes bound to
endoplasmic reticulum

Figure 1.1

A diagrammatic representation of an animal cell showing cross sections of important cell organelles.

Each of the polypeptides is a polymer consisting of a sequence of amino acids arranged in a linear fashion. There are 22 different amino acids, and protein chains can be very different from each other because the amino acids will be arranged in varying sequences. This would even be true with protein chains of the same length. All these amino acids are essential for the growth, development and maintenance of health of an individual. The linear sequence of these amino acids determines the way the chain folds up in the cell and this in turn specifies the shape and conformation of the resulting molecule. These proteins have varying functions in the organism, for example hemoglobin transports oxygen, and enzymes help in performing chemical reactions. Fourteen of the amino acids can be synthesized in the body (nonessential amino acids), but the other eight (essential amino acids) must be obtained from the diet.

Protein is the major building block for muscle, skin, nails, hair, organs, enzymes and antibodies, and is also important as a source of energy. In addition, it is also used in elimination of waste products from the body.

3.   The nucleus

The primary source of instructions for cellular activity and inheritance can be found within the nucleus, and this is our goal for identifying genetic material and our target when studying the chromosomal material. The nucleus is surrounded by a double membrane, as observed by the electron microscope, which appears to be in active contact with the endoplasmic reticulum and the cell membrane. Using the light microscope, a dark-staining network inside the nucleus called chromatin, can be revealed by staining with various chemical dyes, which during cellular division becomes condensed and organized into distinct packages known as chromosomes, the objects of our desires! When the cell is resting, the dark-staining network houses 'resting' chromosomes, not involved in division of the cell and as such are stretched out and tangled, giving the network appearance of the chromatin.

Also observed within the nucleus are rounded bodies called nucleoli attached to specific chromosome regions called nucleolar organizer regions (see *Chapter 3* for nucleolar organizer region, NOR, staining).

3.1   Deoxyribonucleic acid and ribonucleic acid

Nucleic acid is a polymeric compound of high molecular weight, and is important in the release and storage of energy and also the determination and transmission of genetic characteristics. When nucleic acid was first separated from protein, several biochemists, including Levene and Bass (1931), showed that it was made up of units called nucleotides, each of these components being composed of a sugar, a phosphate group and a nitrogen-containing portion. The sugar contained five carbons (ribose, or when lacking an oxygen atom, deoxyribose). Hence there are two kinds of nucleic acid: ribonucleic acid (RNA) and deoxyribonucleic acid (DNA). DNA and RNA are information macromolecules present in almost all cells. The genetic information is stored in DNA which is in turn used to synthesize RNA molecules using sequences of DNA as templates (or molds), which also allows the synthesis of protein polypeptides using RNA molecules as templates.

Molecules of DNA are found in the chromosomes of the nucleus, the mitochondria of animal cells and in the chloroplasts of plant cells.

The more variable component of the nucleotides, the nucleic acid, is the nitrogen-containing group, which contains one or two carbon–nitrogen rings and functions as a base (hydrogen-ion acceptor), in contrast to the acidic phosphate group. Bases containing one carbon–nitrogen ring are called 'pyrimidines', and the two-carbon-ring bases are given the name 'purines'. In DNA, the two main pyrimidines are cytosine and thymine, while RNA carries cytosine and uracil (instead of thymine). The two main purines, adenine and guanine, are found in both DNA and RNA.

Taking the known facts into account, in 1953 Watson and Crick (1953) proposed a double-helix structure for DNA, which quickly became the accepted structural model (*Figure 1.2*). Watson and Crick proposed that the DNA molecule was a double-stranded, double-helix structure, and only by permitting the ends to revolve freely can the two complementary strands be separated. Their circular staircase analogy was artistic and accurate. The

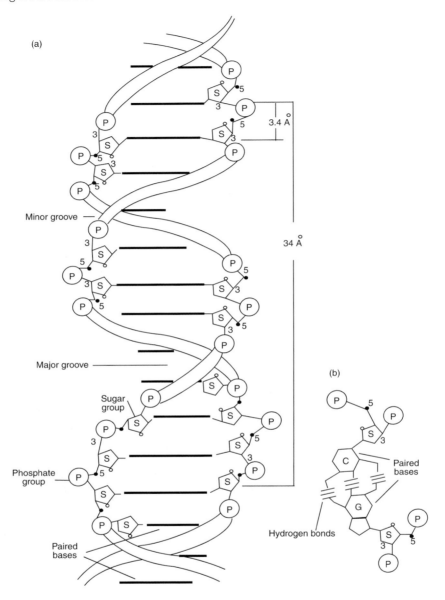

Figure 1.2

Watson and Crick's DNA model: (a) A diagrammatic representation of the double helix with the phosphate sugar backbone of each strand shown by a long ribbon. Between the two strands, parallel rows of paired nucleotide bases are shown, stacked at regular 3.4 Å intervals. Some phosphate (P) and sugar groups (S) are drawn separately. Note the opposite orientations of the sugar groups in each strand. (b) A cross section of the DNA double helix, showing two paired nucleotide bases (C, cytosine, and G, guanine) which extend into the central part of the molecule. The phosphate and sugar groups do not lie in the same flat plane as the paired bases, but are shown in their approximate relative positions.

DNA strands are held together by weak hydrogen bonds, in a DNA duplex, between laterally opposed base pairs, adenine (A) binding to thymine (T), and cytosine (C) to guanine (G). Each turn of the DNA helix has a pitch of 3.4 nm, which accommodates 10 nucleotides. The phosphodiester bonds link the carbon atoms, 3′ and 5′ of successive sugar residues. One end of each DNA strand, the so called 5′ end, will not be linked to a neighboring sugar residue. The other end will be defined as the 3′ end for the same reason. The two DNA strands are said to be anti-parallel because they always anneal in such a way that the 5′ to 3′ direction of one strand is opposite to that of the other.

4.   DNA replication

The process of DNA synthesis (replication) requires that the two strands of DNA are unwound by a helicase (just like the zip of a zipper) and each strand in turn directs the synthesis of a complementary strand. The two daughter DNA duplexes each contain one newly synthesized strand, and hence the replication is described as a semi-conservative. The enzyme DNA polymerase catalyzes the synthesis of new DNA strands using four deoxynucleoside triphosphates (dATP, dCTP, dGTP and dTTP) as nucleotide precursors. DNA replication is time-consuming (not really surprising when you imagine the complexity of the process) – human cells in culture require approximately 8 h to complete the process.

The process of replication is initiated at specific points, called origins of replication. The initiation results in a replication fork where the parent DNA splits to form two daughter duplexes. These daughter duplexes act as templates for the synthesis of complementary DNA strands. The direction of synthesis of one daughter strand is 5′ to 3′, named the leading strand, but 3′ to 5′ for the other, called the lagging strand. Only the leading strand has a 3′ hydroxyl group at the origin of replication, which allows for continuous replication along the entire length of the DNA. However, the synthesis of the lagging strand occurs as a series of small steps, or fragments (usually about 100–1000 nucleotides long). On the lagging strand, the synthesis of DNA strands is called semi-discontinuous. Each fragment on the lagging strand is synthesized in the 5′ to 3′ direction, as illustrated in *Figure 1.3*, which is opposite to the way in which the replication fork moves. The fragments so formed are subsequently joined at their ends by the enzyme DNA ligase to ensure the continued growth of the synthesized strand.

4.1   RNA transcription

DNA specifies the synthesis of RNA, which in turn specifies the synthesis of polypeptides to form proteins. The initial step of RNA synthesis, using a DNA-dependent RNA polymerase, is described as transcription, and this occurs in the nucleus of eukaryotic cells (and to a limited extent in mitochondria and chloroplasts). The second step, called translation (polypeptide synthesis), occurs in ribosomes, large RNA–protein complexes found in cytoplasm, mitochondria and chloroplasts. The RNA molecules that specify the polypeptides are known as messenger RNA (mRNA; see *Figure 1.4*).

Only a very small amount of the DNA in cells is ever transcribed. Depending on their requirements, cells will transcribe different segments of DNA (called transcription units). Equally, only a small proportion of RNA produced by transcription will actually be translated into polypeptides, the reasons being:

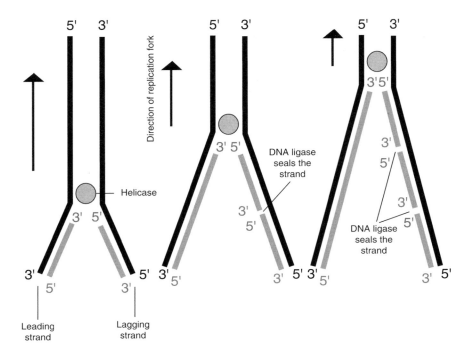

Figure 1.3

DNA strand synthesis during replication: the original strands are shown in the dark color and the newly synthesized strands in the light color. Note that on the leading strand the process is continuous, as the 5′ to 3′ direction of synthesis is the same direction as the fork is moving. However, the lagging strand, the 5′ to 3′ synthesis, is opposite to the replication fork movement, and small fragments are synthesized one at a time and joined by the enzyme DNA ligase. (Figure adapted from Strachan, T. and Read, A.P. (1996) Human Molecular Genetics. BIOS Scientific Publishers, Oxford.)

- the end product of transcription can be an RNA molecule rather than a polypeptide, for example ribosomal RNA;
- the primary transcripts, which encode polypeptides, can be subject to RNA processing which gives rise to smaller mRNA; and
- only a central part of mature mRNA is translated.

RNA synthesis is accomplished using an RNA polymerase enzyme, with DNA as the template and ATP, CTP, GTP and UTP as RNA precursors. The RNA is synthesized as a single strand, the direction of transcription being the 5′ end to the 3′ end. In eukaryotic cells, three different RNA polymerase molecules are required to synthesize the different classes of RNA:

- class I (genes *28SrRNA, 18SrRNA* and *5.8SrRNA*) – localized in the nucleolus, a single primer transcript (*45SrRNA*) is cleaved to give the three ribosomal RNA (rRNA) genes listed;
- class II (all genes that encode polypeptides, and most *snRNA* genes) – polymerase II transcripts are unique in being subject to capping and polyadenylation;
- class III (genes *5SrRNA, tRNA, U6snRNA, 7SL RNA, 7SK RNA* and *7SM RNA*) – the promoter for some genes transcribed by RNA polymerase III is internal to the gene and for others it is located upstream.

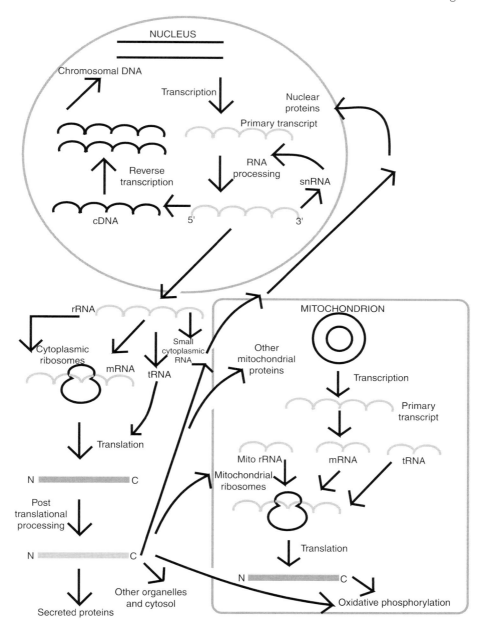

Figure 1.4

Gene expression: a small proportion of nuclear RNA can be converted to cDNA by virally encoded and cellular reverse transcriptases. Mitochondrion synthesizes its own rRNA and tRNA and a few proteins in the oxidative phosphorylation system. (Figure adapted from Strachan, T. and Read, A.P. (1996) Human Molecular Genetics. *BIOS Scientific Publishers, Oxford.)*

5. Mutation and DNA repair

Mutation is a process which produces changes in DNA that may be inherited. It can encompass large changes (including loss, duplication or rearrangement of whole chromosomes or segments – see the section on chromosome organization structure), and point mutations which can affect a single nucleotide. Such changes have the potential to cause disease. Mutations in the germ line, that is the haploid gametes and the diploid cells from which they are formed (see *Sections 6.1 and 6.2*), can be transmitted to subsequent generations. Mutations in the other body cells (somatic cells) result in the disease being confined to the individual in whom the mutation arises, or confined to the type of cells where the mutation originated. The exception to this is when abnormal cells with a mutation of some sort 'break off' and travel via the body's blood or lymphatic system, 'relocating' at another organ or site.

Mutations also offer the possibility of acquiring new improved features for an organism, which may occasionally give it improved reproductive success and by natural selection may confer an advantage over other individuals who do not have the mutation. In these instances, mutations are not necessarily bad for the organism in question. The process of mutation can be increased by ionizing radiation and chemical exposure or occupational/experimental exposure, but the background mutation rate reflects the inevitable errors which occur through chromosome segregation during meiosis (cell division which occurs in the reproductive cells), DNA replication and DNA repair.

A well-known example of this type of mutation giving a selection advantage to an organism is the moth *Biston betularia*. Prior to the Industrial Revolution this moth was naturally a light color, although some darker colored moths were also present. The numbers of the darker moths increased when the grime and pollution caused by the new machinery and smoke from the mills formed a dark background to the environment. Now the dark moth was able to camouflage itself better and not be eaten by predators, and ultimately it survived, whereas its light-colored counterpart became more visible in the smog-like conditions and was detected and picked off, resulting in its effectual demise.

The vast majority of mutations are promptly corrected by DNA repair enzymes that constantly scan the DNA in order to detect and replace damaged nucleotides.

6.   Chromosome organization and structure

Chromosome number and structure vary between species. Bacterial cells typically have a single circular chromosome, but eukaryotes have several large linear chromosomes in their cell nuclei (*Table 1.1*). The chromosome number and DNA content in a complex eukaryotic organism can also vary, for example the gametes (sperm and egg cells) of mammals are specialized sex cells which contain half the number of chromosomes (haploid) of the somatic cells (diploid). The gametes arise from certain somatic cells in the ovary and testis which undergo a specialized form of cell division called meiosis, which serves to reduce the chromosome number by half.

As can be seen in *Table 1.1*, man has a chromosomal number of 46, and in the sperm and egg cells this number is halved to 23, comprising one sex chromosome X or Y, and 22 autosomes (nonsex chromosomes). Diploid cells owe their origin to the fusion of two haploid cells, a process occurring

Table 1.1 Chromosome numbers in different species of animals and plants

Common name	Species	Diploid chromosome number
Carp	*Cyprinus carpio*	104
Dog	*Canis familiaris*	78
Cattle	*Bos taurus*	60
Horse	*Equus calibus*	64
Donkey	*Equus asinus*	62
Silkworm	*Bombyx mori*	56
Tobacco	*Nicotiana tabacum*	48
Potato	*Solanum tuberosum*	48
Man	*Homo sapiens*	46
Rabbit	*Oryctolagus cuniculus*	44
Rhesus monkey	*Macaca mulatta*	42
Rat	*Rattus norvegicus*	42
Mouse	*Mus musculus*	40
Squash	*Cucurbita pepo*	40
Cat	*Felis domesticus*	38
Honeybee	*Apis mellifera*	32
Frog	*Rana pipiens*	26
Rice	*Oryza sativa*	24
Tomato	*Lycopersicon esculentum*	24
Kidney bean	*Phaseolus vulgaris*	22
Onion	*Allium cepa*	16
Barley	*Hordeum vulgare*	14
House fly	*Musca domesticus*	12
Broad bean	*Vicia faba*	12
Fruit fly	*Drosophila melanogaster*	8
Mosquito	*Culex pipiens*	6

during fertilization of the male and female sperm and egg cells to form a zygote (the fertilized egg cell). Male gametes can have either an X or Y chromosome, but female gametes will always have an X chromosome.

The diploid zygote, and all resulting cells in the individual, contains 22 pairs of homologous chromosomes, that is a pair of chromosome 1s, chromosome 2s and so on up to 22, and a pair of sex chromosomes, either XX for a female or XY for a male. Each parent donates one of the homologous pair of chromosomes, that is one chromosome 1 is donated from the female and one chromosome 1 from the male. Diploid somatic cells arise from binary cell division known as mitosis and cytoplasmic division (cytokinesis). During any somatic cell development, and in the mature organism, constituent cells will undergo many rounds of mitotic cell division prior to terminal cell differentiation or cell death. Some mature differentiated cells have no nucleus, for example red blood cells and platelets, and terminally differentiated skin cells lack all organelles. There are also some multi-nucleated cells, such as muscle fibers, which are formed from the aggregation of many cells, and are therefore polyploid (i.e. have multiple numbers of a full haploid set of chromosomes). Additional chromosome sets are present due to extra rounds of DNA duplication prior to division (endomitosis), for example giant megakaryocytes usually contain 16–32 times the haploid DNA content, these in turn giving rise to thousands of platelet cells.

Chromosomes in man consist of two separate arms, divided by a structure called the centromere (see *Figure 1.5*) with the long arm being designated the 'q' arm and the shorter arm the 'p' arm. There are also three distinct shapes – metacentric (where the centromere is central, such as in chromosomes 1 and 19), submetacentric (where the centromere is slightly off-center, such as in chromosomes 4, 6 and 12), and acrocentric (where the centromere is almost at the top of the chromosome such as in chromosomes 13 and 22).

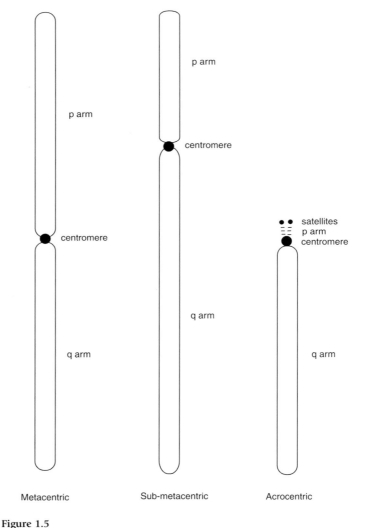

Figure 1.5

Chromosome shapes observed in the chromosomes of man.

6.1   Mitosis

The DNA content changes during the mitotic cell cycle even if the chromosome number remains constant. The cell cycle comprises various phases, of which mitosis is the shortest. Interphase is the process of the cycle where no chromosome condensation is visible, and the interphase is itself divided into three stages, G1 (the period following mitosis until the S phase), the S phase (DNA synthesis phase), and G2 (the period following the S phase prior to mitosis). Nondividing cells remain in a modified G1 phase, sometimes also called G0. The DNA content of a haploid cell is called $1n$, and hence in the cycle of a somatic cell the DNA content may be $2n$ or $4n$, depending upon which part of the cell cycle is being studied. During the S phase, DNA is synthesized and duplicated but the centromere does not divide at this point, and hence the number of chromosomes present remains the same (*Figure 1.6*). The mitotic metaphase chromosome, the one normally seen illustrated and with which we are familiar in the clinical field, consists of two sister chromatids joined together at the centromere.

Stages of mitosis (*Figure 1.7*)

Interphase   This period prepares the cell for division. The nuclear envelope is intact and no chromosomes are visible. Specific chemicals that constitute

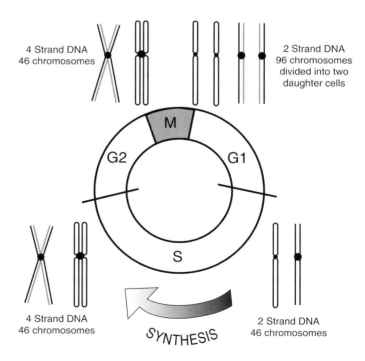

Figure 1.6

Diagram of the cell cycle showing the relationship between the periods of synthesis (S) and mitosis (M). The G1 interval gap can be variable, but the G2 gap between DNA synthesis and cell division and the S phase are relatively constant. Note that the DNA content is doubled during the synthesis phase through to the end of mitosis.

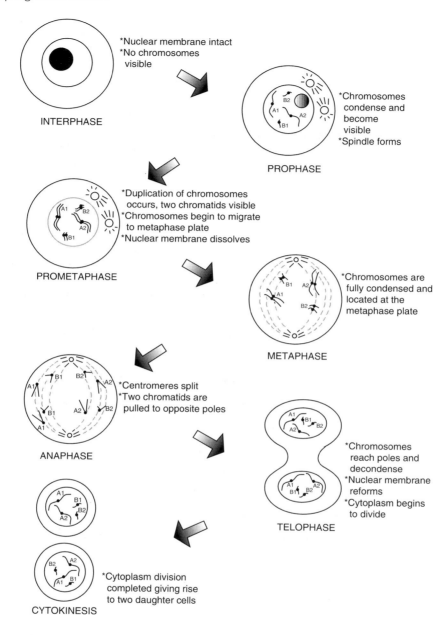

Figure 1.7

Stages of mitosis showing the fate of two homologous chromosome pairs (represented by A1, A2 and B1, B2).

the newly synthesized chromosomes and proteins that will give rise to the mitotic spindle are observed during this period.

Prophase A crucial component during division is the formation of the mitotic spindle, formed from fibers comprising tubulin-based microtubules and associated proteins. There are two important classes of spindle fibers:

the polar fibers (formed in prophase), which extend from the two poles of the cell to the equator; and the kinetochore fibers (formed in prometaphase), attached to the kinetochore, which is a large multiprotein structure attached to the centromere of each chromatid. These spindle fibers extend in the direction of the spindle poles. The chromosomes begin to condense and become visible as thread-like structures. The nucleolus disappears, and also the nuclear membrane begins to break down.

Prometaphase The kinetochore fibers form. The nuclear membrane now dissolves and the chromosomes (condensing all the time) begin to move to the equatorial plane (known as the metaphase plate), and can be seen to contain two chromatids.

Metaphase At this point the chromosomes are at their most condensed and attach to the spindle-shaped structure via their centromeres. The spindle to which the centromeres attach is usually formed from the two centrioles that originated on one side of the nucleus. These centrioles then move to opposite sides of the nucleus and appear to radiate distinctive lines (astral rays). It is this stage that cytogenetic analysis concentrates on, as the chromosomes are at their most condensed and readily recognized following G-banding treatment (see *Chapter 3*).

Anaphase The mutual attachment of the two sister chromatids now ceases via the splitting of the centromere. The two chromatids are pulled to opposite poles of the cell. The appearance is that of active repulsion of the centromeres and dragging of the chromatids along with them.

Telophase The separated chromatids reach the opposite poles and begin to de-condense and become extended as they were in interphase. The nuclear membrane begins to reform around the two sets of daughter chromosomes and the cytoplasm begins to divide.

Cytokinesis Cytoplasm division is completed and gives rise to two daughter cells, with a diploid number of $2n$.

6.2 Meiosis (*Figure 1.8*)

In man, meiosis involves two divisions of specialized diploid cells – primary oocytes (found in the ovary) and primary spermatocytes (found in the testis). Four spermatozoa are produced from the spermatocytes, whereas in females the cytoplasm divides unequally at each stage, with the products of meiosis I being a large secondary oocyte and a small polar body. The secondary oocyte then gives rise to a mature egg cell and another (secondary) polar body.

Stages of meiosis

First meiotic prophase The first division represents a reduction division where members of homologous pairs are separated but not duplicated, that is the numbers are reduced to half (1n). This first prophase is sub-divided into the following five stages:

(i)   Leptotene: the chromosomes appear as long slender threads, with two tightly bound sister chromatids, and bead-like structures (chromomeres) along their length.

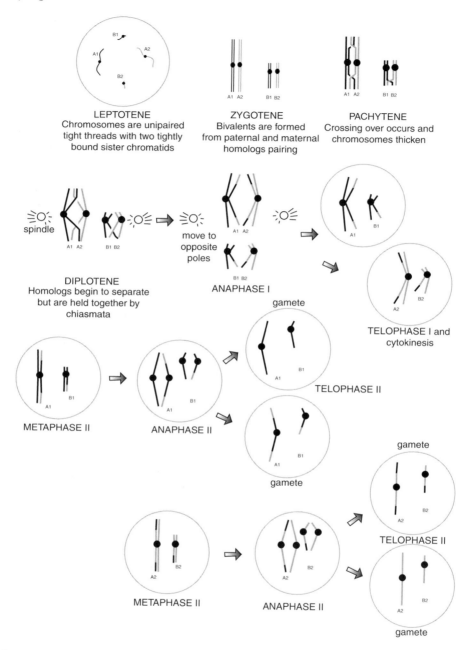

Figure 1.8

Stages of meiosis showing the fate of two homologous chromosome pairs (represented by A1, A2 and B1, B2).

(ii) Zygotene: maternal and paternal chromosomes appear to attract each other and enter a very close pairing (synapsis) where they form bivalents. Each pair of bivalents forms a synaptonemal complex. The pairing is highly specific and occurs between all regions of homologous chromosomes even if they have become translocated (shifted from one chromosome to another) to nonhomologous chromosomes. Similarly if an inversion has occurred (see *Chapter 4*), a loop will form between the two homologs in order to preserve the pairing.

(iii) Pachytene: completion of the formation of synaptonemal complexes marks the start of pachytene, where the chromosomes thicken and condense and recombination (crossover) can occur, which is the exchange of genetic material between nonsister homologous chromosomes. The two homologs can be observed to be physically connected at a specific point called a chiasma (or chiasmata in plural). The chaismata are thought to fulfill an essential function in chromosome segregation, by holding the homologs together until anaphase I, similar to the role performed by the centromeres in mitosis and meiosis II. Each pair of homologs is thought to undergo at least one crossover, but in practice this could be many more, and even the X and Y chromosomes, which exhibit only partial homology, are known to undergo at least one crossover.

(iv) Diplotene: the homologs appear to repulse each other but are still held together by the chiasmata. In many organisms, the position and number of chiasmata seem to be constant for a particular chromosome. At this point the crossovers can be documented and their positions recorded.

(v) Diakinesis: coiling and contraction of the chromosomes occurs until they appear as thick, heavy staining bodies. The bivalents usually migrate close to the nuclear membrane and become evenly distributed. The nucleolus either disappears or detaches from its associated chromosome. During the latter part of this stage, or in the early part of first metaphase, the nuclear membrane dissolves and the bivalents attach themselves by their centromeres to the rapidly forming spindle.

Metaphase I The chromosomes reach a condensed state, as in mitosis, and appear relatively smooth in outline. The chiasmata have moved to the ends of each bivalent (terminalization). These chiasmata prevent the separation of the homologs, which now lie on each side of the equatorial plate of the spindle, stretched by their respective centromeres to opposite poles.

Anaphase I Each chromosome still maintains only a single functional centromere for both of its sister chromatids at this point. This disjunction of either homolog towards opposite poles results in the single centromere dragging both chromatids (dyad) along with it. The chiasmata split off the ends of the chromosomes as they are pulled apart. Each chromosome of a homologous pair is contributed by either parent, and thus the chiasmata exchange during meiosis results in the redistribution of chromosomal material from each parent to both daughter cells formed. The greater the number of chromosomes present, the more the likelihood that redistribution of genetic material from both parents will occur. In each daughter cell there are 23 chromosomes, half the original number.

Telophase I and interphase These stages vary considerably between organisms. In most cases, when the dyad reaches the spindle poles a nuclear membrane is formed around it, and the chromosomes pass into a short interphase before the second meiotic division.

Second meiotic division This is an equatorial division of mitotic type. Metaphase, anaphase and telophase follow, just as in mitosis, but these stages are distinguished from ordinary mitosis by using the subscripts I and II as shown in *Figure 1.8*. The chromosomes enter the prophase of the second meiotic division as dyads, or two sister chromatids, connected together in their centromere regions. As soon as the chromatids divide, each one (monad) separates from its sister and moves to the opposite pole in anaphase II. Telophase II and finally cytokinesis follow rapidly, giving rise to four haploid cells from each initial cell which entered meiosis.

7.   Chromosome function

Chromosomes in mammals have two main functions: (a) perpetuating the hereditary material during an individual's development; and (b) to shuffle and move that material through successive generations.

There are three sequence elements of DNA responsible for the biological functions of eukaryotic chromosomes.

Centromeres These are *cis*-acting DNA elements responsible for segregating the chromosomes at mitosis and meiosis. If chromosomal segments lack a centromere (acentric fragments), they will not become attached to the spindle so will fail to be included in the subsequent formation of daughter cells. A chromosome needs a centromere to join in the fun of division!

Telomeres These seal the ends of the chromosomes and provide chromosome stability. They maintain the structural integrity of the chromosome. If the telomere is lost, the chromosome becomes unstable and tends to fuse with other broken chromosomes if it can. The telomeres also ensure complete replication of the extreme ends of chromosome termini. They also play a role in establishing the three-dimensional structure of the nucleus and chromosome pairing.

Origins of replication DNA in most diploid cells replicates only once per cycle and the control of this initiation of replication is governed by *cis*-acting sequences.

7.1   Chromosome architecture and transcriptional activity

During mitosis, when the chromosomes condense, they are transcriptionally inactive, but during interphase the chromatin fibers are less densely packed and the euchromatin observed dispersed in the nucleus stains diffusely. The remaining chromatin comprises highly condensed fibers and forms dark-staining regions called heterochromatin, and these regions are transcriptionally inert.

There are two types of heterochromatin:

(i)   facultative heterochromatin, which can be genetically active or inactive (i.e. in the special case of the mammalian X chromosome inactivation); and

(ii) constitutive heterochromatin, which is always inactive.

These regions are composed of certain repetitive DNA sequences observed in the centromeric regions of the chromosomes and the p (short) arms of acrocentric chromosomes.

Condensation of chromatin is associated with loss of gene expression, although there is also a large amount of transcriptionally inactive DNA in euchromatic regions. In fact it is known that dark-staining G-bands (see *Chapter 3*, and *Figure 3.1*) contain more condensed chromatin than R-bands, and *in situ* hybridization has also shown that 80% of genes in the human genome map to R-bands and only 20% to G-bands. It is almost like imagining that the chromosome becomes pulled out in interphase but squeezed up like a concertina during the process of division.

7.2 The difference between the X and Y chromosomes of mammals

The major gene that determines maleness is carried on the Y chromosome, which is much smaller than its partner chromosome, the X, a large sub-metacentric chromosome with a large number of genes. However, the presence of the Y ensures maleness, despite any number of X chromosomes, as is the case in Klinefelter syndrome (47,XXY, 48,XXXY or 49,XXXY). However, some XY individuals can be female, where the *SRY* gene located on the Y chromosome is defective in expression or absent. Conversely, XX individuals can be male where a rare translocation translocates the *SRY* gene onto the X chromosome.

The importance of gene dosage is shown in *Chapter 4* regarding gain or loss of chromosomes, for example loss or gain of whole chromosomes is almost always incompatible with life (the X chromosome being an exception). The presence of an extra chromosome usually results in the pregnancy not reaching term and, if it does, severe developmental abnormalities result. Those that have been observed, such as extra copies (trisomy) of chromosomes 13, 18, 21 and X in humans, have been shown to be lacking in transcribed genes compared with other chromosomes.

Both paternally and maternally derived copies of almost all functional autosomal genes are expressed except in the case of imprinted genes, in which case the paternal or maternal copies are expressed (but not both). Because males have one X chromosome and females two, there is an imbalance of the ratio of X-linked genes in the two sexes. To compensate for this difference, one randomly selected X chromosome is inactivated (X inactivation or lyonization). This inactivated X assumes a highly condensed heterochromatic form and is transcriptionally inactive. The inactive X can be observed in interphase by light microscopy as a distinct structure, the Barr body, which is located near the nuclear membrane and replicates late in S phase.

References

Levene, P.A. and Bass, L.W. (1931) *Nucleic Acids*. Chemical Catalog Co., New York.
Watson, J.D. and Crick, F.C. (1953) *Nature* **171**: 737.

Further reading

Strachan, T. and Read, A.P. (1996) *Human Molecular Genetics*. BIOS Scientific Publishers, Oxford.

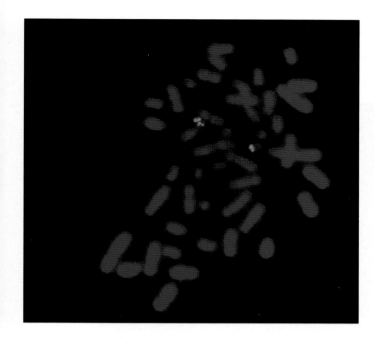

Figure 6.4b

(b) A metaphase spread demonstrating the use of the DiGeorge molecular probe for this deleted region. The probe used for hybridization is by VYSIS with the region 22q11.2 LSI TUPLE 1 shown by the spectrum orange/pink signal and spectrum green indicates the region 22q13 LSI ARSA. Absence of the orange pink signal on one chromosome 22 indicates deletion of the TUPLE 1 locus at 22q11.2. (Photograph supplied by Angela Douglas of the Liverpool Women's Hospital.)

Figure 6.5

Metaphase spread demonstrating the use of the Prader Willi/Angelman molecular probe for detecting the microdeletion at band 15q11-13, the probe used is by VYSIS. The region 15p11.2 D15Z1 is indicated by the spectrum green signal, with two spectrum orange signals, one for the region 15q11-13 D15S10 and 15q22 PML. Absence of the spectrum orange/pink signal on one chromosome 15 indicates a deletion of the D15S10 locus. (Photograph supplied by Angela Douglas of the Liverpool Women's Hospital.)

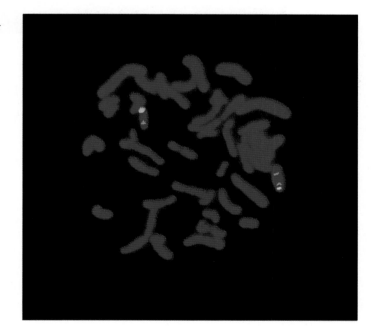

Color section sponsored by Appligene Oncor

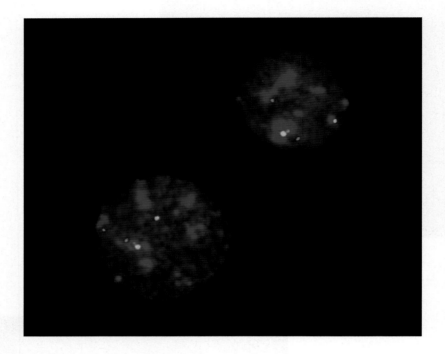

Figure 8.1

(a) Dual color alpha satellite probes for chromosome 6 and chromosome 12 on interphase cells from cultured cell suspension. Chromosome 6 (D6Z1) (product code No. CP5009) biotin labeled and detected with Rhodamine (red). Chromosome 12 (D12Z3) (product code No. CP5031) digoxigenin labeled and detected with FITC (green).

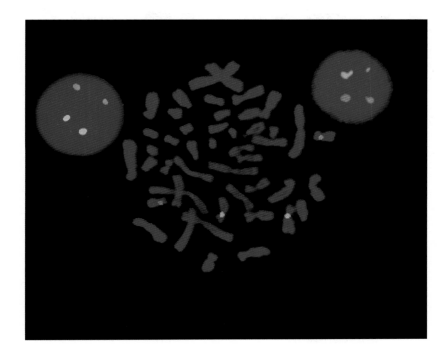

Figure 8.1

(b) Dual color alpha satellite probes for chromosome 8 and chromosome 17 on interphase and metaphase cells from cultured cell suspension. Chromosome 8 (D8Z1)(product code No. CP5013) biotin labeled and detected with FITC (green). Chromosome 17 (D17Z1)(product code No. CP5040) digoxigenin labeled and detected with Rhodamine (red). Note the position of the signal on the metaphase chromosomes at the centromere. (All probes are manufactured and supplied under the label of: Appligene Oncor, Parc d'innovation, Rue Geiler de Kayserberg, BP72, 67402 Illkirch, Cedex France. These photographs were supplied by Roslyn Carruthers of Appligene Oncor).

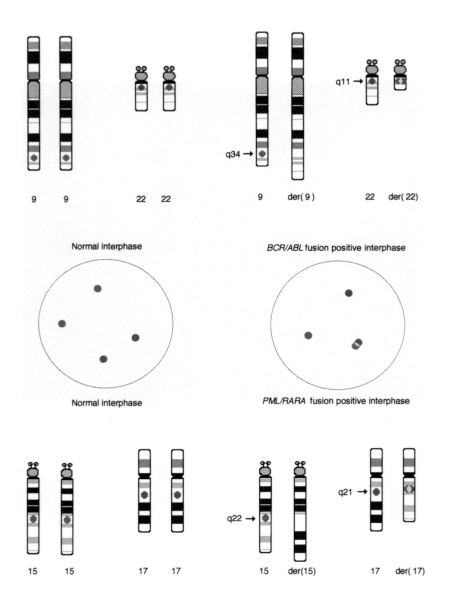

Figure 8.2

(a) Diagrammatic representation of the visual effects, as observed in the interphase cell and on metaphase chr omosomes, of unique sequence probes and their fusion signals. The translocations illustrated are the t(9;22) BCR/ABL fusion and the t(15;17), PML/RARα fusion.

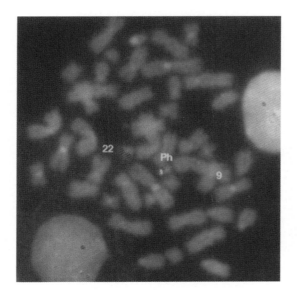

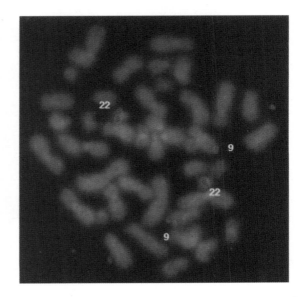

Figure 8.2

(b) BCR/ABL fusion probe (Appligene Oncor) on a t(9;22) translocation positive metaphase. The fusion signal can be seen on the der(22) chromosome (labeled 'Ph' indicating the Philadelphia chromosome), as a 'traffic-light' effect of red, yellow and green. This fusion signal may also be observed in the interphase cells located around the metaphase spread. The same probe is demonstrated on a BCR/ABL fusion negative metaphase.

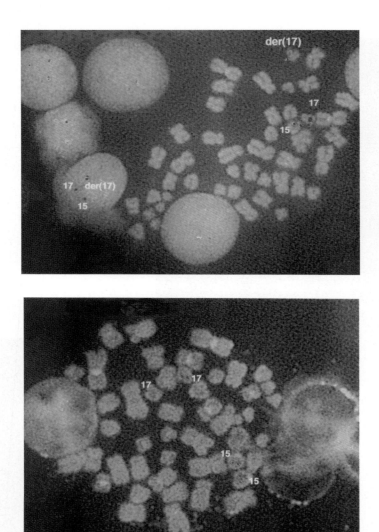

Figure 8.2

(c) PML/RARα fusion probe (Appligene Oncor) on a t(15;17) translocation positive metaphase. The fusion signal is shown on the der(17) as indicated, and may also be observed in the surrounding interphases. The same probe is demonstrated on a PML/RARα fusion negative metaphase.

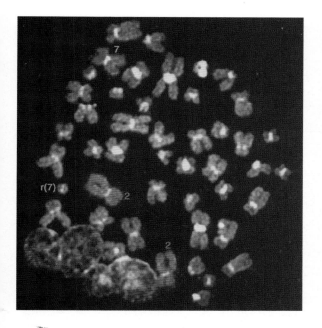

Figure 8.3

(a) Chromosome painting illustrating a normal chromosome 7 and a ring chromosome 7 (green) and two normal copies of chromosome 2 (red).

Figure 8.3

(b) Three-color chromosome painting showing a near tetraploid cell line, chromosome 7 painted in red, chromosome 15 in yellow and chromosome 8 in green. There are four normal copies each of the chromosomes 7 and 15, but the green chromosome 8 paint indicates two normal chromosome 8s and three derivative chromosome 8s, or chromosomes with 8 material. Note also the chromosome paint effects in the interphase cells.

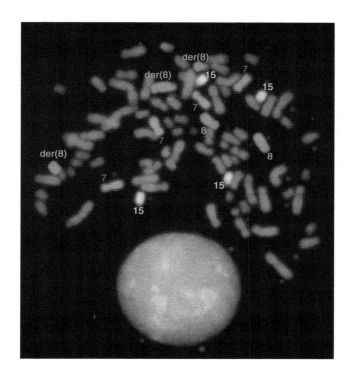

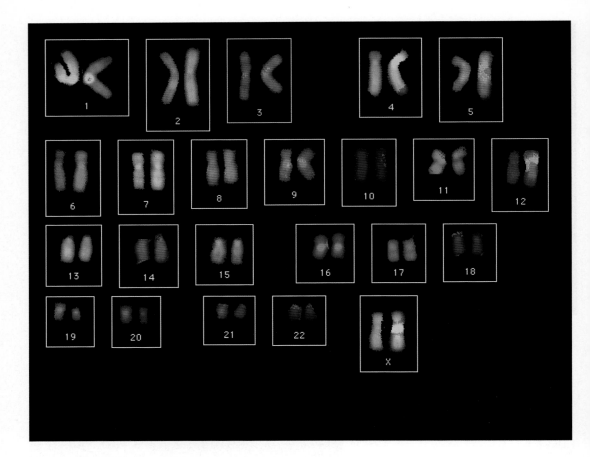

Figure 8.4

Twenty-four color whole chromosome painting and multiplex FISH (M-FISH) analysis of the constitutional karyotype in a patient with unexplained mental retardation. All 24 chromosomes have been pseudocolored according to their unique fluorochrome composition using Powergene M-FISH software (Perceptive Scientific International, Chester, UK). M-FISH revealed a cryptic der(4)t(4;20). The G-banded karyotype was apparently normal. (This photograph was supplied by Dr Lyndal Kearney of the Institute of Molecular Medicine, John Radcliffe Hospital, Oxford).

Chapter 2

Culture of human cells for chromosomal analysis

Karen Saunders and Barbara Czepulkowski

I. Introduction

Only cells that are actively dividing can be used for cytogenetic studies. All samples for analysis should, therefore, arrive as fresh as possible at the laboratory having spent a minimum time in transit. Cells will survive for a few days if refrigerated, but they will die if frozen or exposed to high temperatures.

Every cytogenetic laboratory has a slightly different method of culturing samples for analysis. Techniques that work for one laboratory may not necessarily work as effectively for another. Many factors influence cell culture, ranging from local variations, including temperature and humidity, to differences in the companies used to provide laboratory supplies and the way that samples are handled by the referring clinicians and hospitals. However, the techniques used in each laboratory are all based on, and incorporate, the same common principles.

2. Culture of peripheral blood samples

2.1 Receipt of the sample

In order to successfully culture blood samples for cytogenetic analysis they have to be received fresh and unclotted. The samples are therefore collected in sterile tubes containing the anti-coagulants sodium or lithium heparin. Any other choice of tube is unsuitable since the sample will either arrive clotted or the tube will contain a substance such as EDTA, which is toxic to the cells and will, therefore, adversely affect the culture.

2.2 Blood cell culture introduction

Normally only the white blood cells are used for cytogenetic analysis (i.e. the neutrophils, eosinophils, basophils, monocytes and lymphocytes). These cells contain a nucleus and are capable of undergoing cell division. T lymphocytes are stimulated into the process of cell division when exposed to certain antigens, and these cells are easily available from the patient/individual being studied. The mucoprotein phytohemagglutinin (PHA), found naturally in the red kidney bean (*Phaseolus vulgaris*), is a mitogen commonly used in the cytogenetics laboratory. It behaves in a similar way to a foreign antigen, stimulating lymphocytes to divide. By adding PHA to blood cultures, cell division is stimulated by causing an increase in DNA synthesis in the T lymphocytes. The first mitoses are usually observed 48 h after addition of PHA, although when studying fetal or newborn blood a few mitoses may be obtained after 24 h. Mitotic activity then occurs at 24 h intervals, that is 48, 72 and 96 h, and this corresponds to the first, second and third cycles of cell division. Other mitotic agents (mitogens) exist, such as pokeweed mitogen (PWM), however, PHA normally produces the best results in routine blood culture. Pokeweed mitogen induces T cell activity, although it is more effective at stimulating the B cells.

The culture medium used consists of a pH and osmotically balanced salt solution that contains all the nutrients and growth factors required for short-term blood lymphocyte cell culture. The broad-spectrum antibiotics penicillin and streptomycin are added to this to prevent the growth of microorganisms. The presence of such organisms may have an adverse effect on the growth of cells. Thymidine is then added to the blood cultures to produce a chemical imbalance that blocks DNA synthesis at a certain point of the cell cycle. After addition of thymidine, the cultures are re-incubated for a further 16 h, when many of the cells will be entering cell division, but will be blocked from continuing further.

Finally, colcemid is added to inhibit formation of the spindle, thereby introducing a second block between metaphase and anaphase. This is when the chromosomes are best visualized. The longer the treatment with colcemid, the more metaphase chromosome spreads are obtained. However, prolonged treatment results in a reduction of chromosome length since the chromosomes continue to condense with time. By synchronizing cell division with thymidine, colcemid treatment can be reduced to a minimum time of about 20 min. This allows the harvest of a relatively high proportion of early metaphase cells and thereby produces longer, better quality chromosomes. The next stage in processing blood samples is addition of a hypotonic solution, usually potassium chloride. This swells the cells and so facilitates chromosome spreading during slide-making. Prolonged treatment with KCl is not required for routine blood culture and cultures can be centrifuged immediately and the supernatant removed. Fixative, a mixture of methanol and acetic acid in a proportion of 3:1, is then added to cultures, which ensures the removal of cytoplasm and cell debris. The components of the fixative work in different ways. The acetic acid precipitates nucleic acids, therefore the chromosomes are fixed but the surrounding material is dissolved or softened, whereas the methanol precipitates proteins. By mixing methanol and acetic acid in the above proportions, an optimum balance between the best fixation of the chromosomes and removal of surrounding cellular material is achieved.

3.   Culture of cells from amniotic fluid

3.1 Receipt of the sample

The cells in amniotic fluid are derived from the amnion and fetal tissues such as the skin and urinary tract. Using a clinical procedure termed amniocentesis, a sample of the amniotic fluid is collected, preferably into two sterile Universal containers. Between 15 and 20 ml, depending on gestation, is required for cytogenetic analysis. It is particularly important that the samples should arrive in the laboratory as soon as possible following amniocentesis since the cells are very vulnerable to deterioration. On receipt the sample is booked into the laboratory and given an identification number.

3.2   Introduction to amniocyte cell culture

Amniocytes are grown in culture vessels, the cells divide and grow as a single layer attached to the culture vessel surface producing colonies of cells. Each colony originates from one parent amniocyte cell. All culture work is carried out in sterile conditions in a class II safety cabinet.

3.3   Cell harvest

Once adequate cell growth is observed the culture may be harvested. There are two main harvesting techniques.

In situ

The cells are harvested and examined on the surface on which they were originally seeded, (in situ). This technique is used for cultures that have been grown onto either coverslips within a culture vessel or slide flasks. The cultures are ready for harvest when sufficient colonies with adequate cell growth are observed and plenty of cells are entering mitosis. The mitotic cells can be easily recognized since they are rounded in appearance, as opposed to the flattened appearance of nondividing cells, and situated mostly around the periphery of the colonies.

Suspension

These cultures are ready for harvest when they show almost confluent growth across the culture vessel surface with many dividing cells. The cells are removed from the original culture vessel, after mitotic arrest with colcemid, by treatment with trypsin, and harvested in suspension. The fixed cell suspension is then dropped onto microscope slides for examination. The culture time for this method is therefore slightly longer than that for the in situ method, but has the advantage that more divisions can be obtained for examination.

3.4   Alternative harvest techniques

Thymidine blocking

Some laboratories use the addition of chemicals such as thymidine to the cultures the night before harvest to block all the dividing cells at a certain point of the cell cycle. Release from the block the following morning then results in a synchronization of cell division. More cells are therefore reaching metaphase at the same time and subsequent colcemid treatment can be reduced resulting in longer chromosomes.

Overnight addition of bromodeoxyuridine with colcemid

Chemicals that inhibit chromosome condensation, such as bromodeoxyuridine (BrdU), can be added together with colcemid for a prolonged overnight colcemid treatment, which also results in good-quality chromosome preparations (see Protocol 2.5).

4.   Culture of chorionic villus samples

4.1   Receipt of the sample

A sample of chorionic villi (CVS) for cytogenetics is taken from the chorion frondosum at around 12 weeks gestation. The sample obtained is usually about 20 mg in size. The sample should be sent directly to the laboratory in transport medium, consisting of a basal medium, fetal calf serum, the broad-spectrum antibiotics penicillin and streptomycin, fungicide (e.g. nystatin) and lithium heparin. The heparin prevents the clotting of any maternal blood in the sample, which would hinder the sorting and cleaning of the sample.

4.2   Introduction to CVS culture

Two different cultures can be obtained from a CVS sample, a direct culture and a long-term culture. The direct culture relies on the spontaneously dividing cells in the cytotrophoblast layer of these chorionic villi in order to produce a result, and subsequently the cultures are processed after only a very short time in culture medium. The quality of the preparations tends to be poorer than those obtained from the long-term cultures and they are only used to provide a rapid initial aneuploidy screen. The long-term culture relies on the dividing cells in the mesenchyme core of the CVS to produce a result. The cells are grown in culture vessels, in a similar way to amniocytes, as a single layer attached to the culture vessel surface.

5.   Solid tissue culture

5.1   Receipt of the sample

The tissue must always be sent to the laboratory unfixed and, for smaller samples, in an appropriate transport medium, such as an osmotically balanced salt solution containing antibiotics and fungicides.

5.2   Introduction to solid tissue cell culture

Samples suitable for cytogenetics include:

■ Skin – small skin biopsies from live patients or a post-mortem sample.
■ Post mortem fetuses – fetuses from terminated or spontaneously aborted pregnancies. Small samples of skin and muscle are usually removed for cytogenetic purposes.
■ Products of conception (POC) – this includes the contents of the uterus after early termination or miscarriage. The sample is sorted to remove maternal decidua and blood and chorionic villus is taken for culture if available.

5.3   Setting up of cultures

The cultures are set up and harvested in exactly the same way as the long-term cultures for CVS samples (*Protocol 2.8B*). The only variation may be the duration of exposure of the tissue to the enzymes used in the enzymatic degradation method. Different tissues may need a slightly longer or shorter exposure. This also depends on the degree of natural maceration of the tissue, which may have already occurred due to deterioration of the sample before collection. For example, tissues from fetuses that have died *in utero* have often deteriorated considerably by the time they can be collected. Some laboratories, therefore, may only use the maceration technique to set up solid tissue cultures.

6.   Culture of bone marrow and leukemic blood

6.1   Receipt of the sample

Bone marrow must be heparinized immediately upon aspiration since clotting renders the sample useless for cytogenetic study. Thus the sample should be transported in a sterile container containing one of the following according to preference:

■ 0.3 ml of preservative-free heparin (stock solution 1000 U ml⁻¹);
■ 2.0 ml Hank's balanced salt solution containing 0.3 ml of heparin as above;
■ 5 ml transport medium (made up from 100 ml basal medium, e.g. McCoys 5A, 1 ml of penicillin 10 000 IU ml⁻¹, 1 ml of streptomycin 10 000 μg ml⁻¹ and 1 ml preservative-free heparin 1000 U ml⁻¹).

It is preferable that the samples are received as soon as possible. If a delay is anticipated, such as that expected when the referring institution is some distance away from the laboratory, then the use of transport medium is strongly recommended to minimize drying out of the sample and to maintain cell viability.

Inevitably there will be occasions when samples arrive at inconvenient times for the laboratory, or when some delay occurs in dispatch of samples. In these circumstances, samples should be stored at 4°C overnight and for no longer than 3 days.

The sample can deteriorate markedly following any sort of delay in dispatch, leading to a possible misleading result if normal cells have outgrown the abnormal cells. Samples from patients with high white cell counts are subject to deterioration as the high number of cells cannot be maintained sufficiently by the transport medium. Consequently, any delay with this type of sample can prove costly, and any subsequent failure is also hard to explain to the referring hospital. Setting up the culture with the minimum delay is obviously the preferred scenario.

6.2   Introduction to bone marrow, leukemic blood, and lymph node cell culture

Patients suffering from malignant disorders are often immunocompromised, and thus particularly susceptible to pathological infection. This fact makes these samples potentially hazardous to the health and safety of the laboratory staff. Specimens received from all patients being investigated for hematological and malignant disorders must therefore be handled in a class I microbiological safety cabinet, using full aseptic technique.

The tissue of choice for the cytogenetic study of most hematological conditions is the bone marrow, where the source of immature abnormal cells is usually found. Exceptions to this are chronic granulocytic leukemia (CGL) and chronic lymphocytic leukemia (CLL), as mature cells are often involved in these cases and, as such, a blood sample is more appropriate. Also, because of the high white counts encountered in CGL and CLL, it is sometimes physically not possible to aspirate bone marrow. Even when this is obtainable, blood may be more informative, as there are fewer cells (abnormal and normal) crowding and complicating the processing of the material. When studying lymphomas, the bone marrow may not always be involved in the early stage of the disease, and as such a lymph node biopsy is usually more informative. Generally, by the time the bone marrow has become involved in high-grade lymphomas, there may be many chromosomal changes present, and this makes it difficult to establish the important primary change.

Bone marrow aspirates from the sternum or posterior iliac crest are usually successful, but occasionally a dry tap may occur, which can be due to the following:

■ myelofibrosis or osteosclerosis in myelodysplasia;
■ secondary myelofibrosis or osteosclerosis in tumors;
■ compact cellular marrow;
■ reticulo endotheliosis;
■ faulty technique.

In these cases a peripheral blood sample can be sent to the laboratory as an alternative. It should be noted that a blood sample is only of use if sufficient blast cells are present in order to detect an abnormal clone.

A sample size of 1–2 ml is adequate in most cases. It is advisable to ensure that the sample is not the final exudate from the syringe, as the initial tap may contain a large number of erythrocytes.

6.3   Setting up tissues from malignant conditions

Bone marrow

A number of approaches will yield metaphase spreads from the above tissues. An unsynchronized set of cultures and a synchronized technique similar to that employed for lymphocyte culture (*Protocol 2.1*) are employed. Unsynchronized culture methods include:

■ direct culture;
■ short-term culture (usually 24 h);
■ an overnight with colcemid culture.

They are essentially similar cultures, but colcemid treatment occurs at different times, and different exposure times.

The blocking agents for synchronized cultures shown below have been found to be effective with bone marrow and also leukemic blood:

■ methotrexate (MTX);
■ fluorodeoxyuridine (FdU);
■ fluorodeoxyuridine and BrdU used in a mixture.

The selection of culture regimes is a matter of preference, but also occasionally has to be tailored to the disease being investigated. This is particularly important in the case of known or suspected acute promyelocytic leukemia (APML) where the diagnostic t(15;17) translocation may only be detected in cultures of 24 h or longer and is not seen in direct cultures.

Peripheral blood

All protocols used for bone marrow samples may be applied to unstimulated peripheral blood cultures. However, mitogens can be added to stimulate the proliferation of lymphocytes in the same way as described in *Protocol 2.1*. When lymphoid diseases are known or suspected to be of B cell type, mitogens are employed using in particular 12-0 tetra-decanoylphorbol-13-acetate (TPA), the B cell mitogen, and pokeweed mitogen (PWM), a B cell and T cell mitogen. When T cell disease is suspected, PHA can also be employed.

Synchronization with a BrdU/FdU cocktail

Protocol 2.10 describes our adaptation of a BrdU synchronization method using a cocktail of BrdU and FdU. Of all the synchronization methods available, we find that the best results are obtained with the method given in *Protocol 2.10*.

7.   Culture of samples from chronic lymphoid leukemia and the lymphomas

7.1   Receipt of the sample

Biopsies of affected lymph nodes in lymphomas are the best tissue to use when studying these diseases. Chronic lymphoid leukemias arise in the more mature blood cells, and as such the best tissue for study of these conditions is peripheral blood. Bone marrow, however, can also be processed.

If the operating theatre is nearby, it is advantageous to place lymph node biopsies directly into the RPMI culture medium given in *Protocol 2.12*, but if this is not the case, transport medium such as that shown in *Section 6.1* can be used. It is best to process the sample as soon as possible to maintain cell viability. Lymph nodes deteriorate rapidly with delays in transit.

7.2   Culture of tissues from chronic lymphoid leukemias and lymphomas – introduction

Chronic lymphoid leukemias are extremely difficult to analyze cytogenetically due to the low spontaneous mitotic index and poor response to most mitogens. As mentioned previously, it is a common problem to be unable to obtain metaphases even from patients with a high white cell count. Ninety percent of all CLLs are of B cell type, and as such B cell mitogens are used to stimulate these abnormal cells. The most commonly used mitogens are TPA (also known as 4-phorbol-12-myristate-13-acetate, PMA), and Epstein–Barr virus (EBV).

Malignant lymphomas are a group of localized tumors with an abnormal proliferation of lymphoid cells, usually arising in the lymph nodes, thymus spleen or mucosal-associated lymphoid tissues, but can occasionally arise in bone marrow, liver, skin or the intestinal tract.

8.   Slide preparation

There are many different methods of preparing slides and each laboratory has a favorite technique. The aim is to produce well-spread chromosomes, which remain as intact metaphase spreads. Any shortcomings in technique will either produce chromosomes which overlap with one another in a tangle (underspreading), or metaphase spreads containing less than the complete modal number (overspreading). An extreme case of overspreading will appear on the slide as 'chromosome soup' with chromosomes having been scattered across the slide.

Using a phase contrast microscope, the quality of a slide can be assessed immediately. If the metaphase cells are not spread well, or the slide is too crowded with cells, for instance, adjustments can then be made prior to making further slides. It is wise to check the slides one by one, even though it may be laborious; this proves worthwhile in the long run as good preparations are essential for ease of subsequent staining procedures.

The chromosomes in the metaphase spreads examined under phase contrast should appear a pale gray color, not black and shiny or pale and ghostly. If there is a layer of cytoplasmic material around the metaphase, which appears like a gray veil covering the chromosomes, G-banding becomes problematical and good crisp clear bands are not always possible.

The rate of fixative evaporation depends upon the atmospheric conditions when the cell suspension is dropped onto the slide. When dropping the suspension onto the slide, one can observe the speed of evaporation by the rate at which the drop recedes and begins to dry on the slide. The cell membrane becomes stretched during this process, and starts to flatten, capturing and fixing the chromosomes onto the slide surface. However, if evaporation is too rapid, which can be observed by the drop of suspension receding and drying too quickly, the cell membrane does not stretch adequately, leaving the problem of the layer of cytoplasm on a poorly spread metaphase. Further hints and tips on spreading are to be found in *Protocol 2.14*.

With practice, spreading stimulated blood chromosomes is relatively easy. Bone marrow chromosomes and occasionally unstimulated blood cultures, however, can be difficult to spread. *Protocol 2.14* outlines a method which should help to produce reasonable slides even in the most stubborn of cases.

Further reading

Rooney, D.E. and Czepulkowski, B. (eds) (1992) *Human Cytogenetics: A Practical Approach*. IRL Press, Oxford.

Protocol 2.1

Setting up cell cultures and harvesting

Equipment

Airflow class II safety cabinet

Centrifuge set at 200 *g* with swinging bucket rotor

Culture vessels (Sterile Universals, or flat-bottomed tubes)

Gassed incubator set at 37°C providing an environment of 5% CO_2, 95% air

Solutions

Colcemid 10 μg ml^{-1}

Complete culture medium made up as follows:

- 100 ml RPMI 1640 basal medium
- 10% fetal calf serum
- 1% phytohemagglutinin
- 1% penicillin
- 1% streptomycin
- 1% L-glutamine

Fixative made up of three parts Analar-grade methanol to one part Analar-grade glacial acetic acid

Potassium chloride 0.75 Mm (5.5 g l^{-1})

Thymidine 15 mg ml^{-1} made up in Dulbecco's phosphate-buffered solution (PBS)

Protocol

1. Add 2.5 ml of whole blood to 5 ml of culture medium, in a sterile plastic tube in the safety cabinet.

2. Incubate the blood cultures at 37°C for 48 h.

3. Add thymidine to the cultures and re-incubate for a further 16 h.

4. Release the thymidine block by centrifuging the sample at 200 *g* for 10 min.

5. Remove the supernatant and replace with fresh medium, thereby removing the thymidine. Now that the cultures are released from the block the cells proceed to divide in synchrony.

6. Re-incubate for 4.5 h then add colcemid to the cultures for 20 min.

7. Centrifuge the sample at 200 *g* for 10 min. Remove the supernatant and replace with 5 ml of the hypotonic solution, potassium chloride.

8. Add a few drops of fixative to the pellet. Thorough mixing is important during addition of the fixative in order to ensure that the cytoplasm is effectively removed.

9. Re-centrifuge the culture tubes and change the fix several times until a 'clean' cell pellet is achieved.

10. Dilute the cells with an appropriate amount of fixative.

11. Make slides as in *Protocol 2.14.*

Hints and tips

There is always the occasional sample that fails despite all attempts made to produce chromosomes. Certain patients are more likely to have a problematic sample due to their conditions. These include:

■ pregnant women who have recently had a miscarriage (or in particular an intra-uterine death);

■ patients who have been prescribed drugs which affect cell growth;

■ neonates with certain conditions or syndromes, for example Di-George syndrome (where the patient lacks T lymphocytes).

Protocol 2.2

Amniotic fluid culture

Equipment

Airflow class II safety cabinet

Centrifuge set at 200 *g* with swinging bucket rotor

Culture vessels, flat-bottomed tubes, flasks or coverslip slide flasks

Gassed incubator set at 37°C providing an environment of 5% CO_2, 95% air

Inverted microscope

Solutions

Complete culture medium, made up as follows:

- 100 ml Amniomax basal medium
- Amniomax supplement, as supplied
- 1 ml 10 000 U ml^{-1} penicillin
- 1 ml 10 000 μg ml^{-1} streptomycin
- 1 ml 200 mM L-glutamine

Protocol

1. Separate the amniocytes from the amniotic fluid by centrifuging the sample at 200 *g* for 10 min and removing the supernatant to leave the pellet of cells at the bottom of the tube.

2. Re-suspend the cell pellet in a small amount of culture medium.

3. Divide the cell suspension between a minimum of two culture vessels so that the suspension just forms a puddle at the bottom. This allows the cells to settle down and attach to the culture surface more quickly. The choice of vessel used varies between laboratories, but the most popular include 35 mm culture dishes containing a coverslip, 45 × 22 mm slide flasks (a microscope slide incorporated into a culture flask), and 25 cm^2 flasks or flat-bottomed tubes (Leighton tubes).

4. Place the cultures into the incubator.

5. Top up the cultures the following day with fresh medium so that it comes up to about 5 mm high in the culture vessel.

6. Leave the cultures to grow undisturbed for 5 days.

7. After this time, remove the medium in the cultures and replace with fresh medium.

8. Check the cultures under an inverted microscope for cell growth.

Protocol 2.3

Harvesting amniocytes

Equipment

Absorbent paper

Airflow class II safety cabinet

Centrifuge set at 200 *g* with swinging bucket rotor

Coplin jar (for slide flask harvest)

Inverted microscope

Syringes (plastic, I ml without a needle)

Solutions

Colcemid 10 μg ml^{-1}

Fixative made from three parts Analar-grade methanol to one part Analar-grade glacial acetic acid

Potassium chloride 0.60 mM

Trypsin solution, 0.05% trypsin, 0.53 mM EDTA

Versene solution (1:5 000 concentration)

XAM or DPX mountant

Protocol A: **In situ** *harvest*

1. Add I drop of colcemid solution from a I ml syringe without a needle per 0.5 ml of medium in the culture vessel and re-incubate for between 20 min and 2 h.

2. Remove the medium and replace with 2 ml of the KCl (or if slide flasks are being used detach the slide from the rest of the vessel and place into a Coplin jar full of the KCl solution).

3. Add fixative very gradually to the KCl, starting with one drop and working up to 0.5 ml over a time period of 10 min, agitating the tube constantly.

4. Replace the fixative/KCl mixture with fresh fixative.

5. Leave for a few minutes, and then replace with fresh fixative twice more.

6. Remove the coverslip or slide from the fixative and tip against some absorbent paper to allow some of the fixative to drain off.

7. Lay it flat on some absorbent paper and allow to air dry.

8. Mount the coverslip onto the slide using XAM or DPX mountant.

Protocol B: Suspension harvest

1. Add 1 drop of colcemid solution per 0.5 ml of medium in the culture vessel and re-incubate for up to 2 h.

2. Remove the medium from the culture vessel and wash with versene solution to prevent the serum in the medium inhibiting the action of the trypsin.

3. Remove the versene solution.

4. Add the trypsin solution and incubate.

5. After 1–2 min check to see if the cells are detached from the surface of the culture vessel using an inverted microscope.

6. When most of the cells are floating freely, add the culture medium. Transfer the cell suspension to a centrifuge tube and centrifuge for 5 min.

7. Remove the supernatant and replace with KCl solution.

8. Incubate for 20 min.

9. Add 2 ml of fresh fixative to the KCl and re-centrifuge for 10 min.

10. Remove about half of the fixative/KCl solution, resuspend the cells and top up with fixative, then re-centrifuge for 10 min.

11. Remove the supernatant and slowly add fixative, taking care to agitate the mixture constantly, and then re-centrifuge for 10 min.

12. Repeat step 11 twice more.

13. Remove the supernatant and resuspend in a small amount of fixative until the suspension appears slightly cloudy.

14. Drop the appropriate concentration of cell suspension onto a microscope slide, as in *Protocol 2.14*.

Hints and tips

Green or black amniotic fluid samples, which are caused by the presence of old maternal blood in the sample from bleeding earlier in pregnancy, sometimes grow poorly and are more prone to culture failure.

If cells are not detaching readily following trypsin treatment, tap the culture vessel with the palm of the hand.

Protocol 2.4

Thymidine blocking

Equipment

As for *Protocol 2.3*

Pasteur pipettes

Solutions

Colcemid 10 µg ml^{-1}

Thymidine 8 mg ml^{-1} in Dulbecco's PBS

Method

1. Add one drop of thymidine solution from a Pasteur pipette for each 1 ml of medium in the culture vessel the night before harvest.

2. Release from the block 16 h later by removing the medium with a Pasteur pipette and replacing with fresh medium.

3. Re-incubate for 4.5 h.

4. Add 1 drop of colcemid solution per 0.5 ml of medium in the culture vessel.

5. Re-incubate for 20 min.

6. Continue harvest as in *Protocol 2.3* from step 2.

Protocol 2.5

Addition of BrdU with colcemid

Equipment

As for *Protocol 2.3*

Solutions

BrdU 3 mg ml^{-1}

Colcemid 1 µg ml^{-1}

Protocol

1. Add 25 µl each of BrdU and colcemid solution per 1 ml of medium the night before harvest.

2. Incubate for 16 h.

3. Remove the medium and continue the harvest with addition of KCl as in *Protocol 2.3* from step 2.

(Protocol provided by Labor. MCL, Duedingen, Switzerland.)

Protocol 2.6

Sorting of the chorionic villus sample

Equipment

Airflow class II safety cabinet

Inverted microscope

Sterile forceps

Sterile plastic Petri dish

Solutions

Culture medium, made up as follows:

- Chang B basal medium 100 ml
- Chang A supplement, as provided
- penicillin, 1 ml (10 000 U μl^{-1})
- streptomycin, 1 ml (10 000 μg ml^{-1})
- L-glutamine, 1 ml (200 mM)
- nystatin, 0.1 ml (1000 μg m^{-1})

Protocol

1. Clean the CVS of maternal blood and mucus by washing in several changes of culture medium in a sterile plastic Petri dish.

2. Sort the CVS under an inverted microscope using sterile forceps by separating the chorionic villi from any maternal tissue and small blood clots into a separate sterile Petri dish containing fresh medium.

Hints and tips

It is extremely important that all maternal tissue is removed from the sample since maternal cells can also grow in culture and may provide a maternal as opposed to a fetal karyotype result.

Protocol 2.7

Direct culture of chorionic villus samples

Equipment

Airflow class II safety cabinet

Centrifuge set at 200 *g* with swinging bucket rotor

Centrifuge tube, sterile

Gassed incubator set at 37°C providing an environment of 5% CO_2, 95% air

Hotplate at 40°C

Inverted microscope

Pasteur pipette, plastic

Pipette, glass

Solutions

Acetic acid 60%

Chang culture medium, made up as in *Protocol 2.6*

Fixative made up of three parts Analar-grade methanol to one part Analar-grade glacial acetic acid

Tri-sodium citrate hypotonic solution, 1%

Protocol

1. Place one-third of the sorted CVS into a plastic sterile centrifuge tube containing Chang medium and incubate for 3 h.

2. Add colcemid for 1.5 h (if the sample is received late in the day then it can be incubated overnight and colcemid added the next morning).

3. Remove the medium using a pipette and replace with 5 ml of hypotonic solution.

4. Re-incubate for 5–10 min.

5. Slowly fix the sample by progressively removing the hypotonic solution and replacing it with fresh fixative.

6. Remove all the fixative and add a volume of 60% acetic acid approximately equal to the amount of tissue present.

7. Carefully aspirate the sample using a pipette until the suspension becomes cloudy in appearance.

8. Add a couple of drops of the suspension to a pre-warmed microscope slide on a hot plate at about 40°C.

9. Slowly, and very gently, spread the suspension back and forward across the microscope slide, using the length of a glass pipette, until most of the fixative has evaporated and only small drops remain.

10. Leave the slide to dry completely.

11. Make as many slides as the suspension will allow for.

12. Allow the slides to age for a few hours on the hotplate before banding.

Protocol 2.8

Long-term culture of chorionic villus samples

There are two main approaches to setting up CVS for long-term culture:

(i) Dissociation using enzymes – two enzymes, trypsin and collagenase are used to digest the sample to a cell suspension, which is then cultured. This method usually results in relatively rapid cell growth.

(ii) Chopping finely with a scalpel (maceration) – the CVS is chopped in a sterile plastic Petri dish using a scalpel into very small pieces and cultured.

Most laboratories use both techniques to provide two different sets of cultures.

Protocol A: Enzymatic dissociation
Equipment

Airflow class II safety cabinet

Centrifuge set at 200 *g* with swinging bucket rotor

Culture vessels, flasks

Gassed incubator set at 37°C providing environment of 5% CO_2, 95% air

Inverted microscope

Solutions

Bacto trypsin, 5% solution

Chang culture medium made up as in *Protocol 2.6*

Colcemid 10 μg ml^{-1}

Collagenase type II 1 mg ml^{-1} in PBS solution

Fixative made with three parts Analar-grade methanol to one part Analar-grade acetic acid

PBS

Potassium chloride 0.60 mM

Versene solution (1:5 000 concentration)

Protocol

1. Wash the CVS sample in PBS solution.

2. Transfer them to a centrifuge tube and incubate at 37°C in trypsin solution for about 1–2 min.

3. Centrifuge the partially digested sample and remove the supernatant.

4. Re-suspend in collagenase solution and re-incubate.

5. Agitate the sample about every 20 min using a pipette to help break the tissue up.

6. Add complete culture medium once the tissue is completely dissociated to prevent further enzymatic action.

7. Centrifuge the sample and discard the supernatant.

8. Re-suspend the pellet in fresh culture medium.

9. Repeat steps 7 and 8.

10. Seed the cell suspension into at least two culture vessels and check and harvest using the same procedure as for amniotic fluid in situ culture in *Protocol 2.3*.

Protocol B: Maceration technique
Equipment

As above, plus:

- Coverslips
- Culture vessels (sterile) – four-well Petri dish
- Scalpel blade (sterile)

Solutions

Colcemid 10 μg ml^{-1}

Chang culture medium made up as in *Protocol 2.6*

Fixative made up of three parts Analar-grade methanol to one part Analar-grade acetic acid

Potassium chloride 0.60 mM

Trypsin-EDTA solution (0.05% trypsin, 0.53 mM EDTA)

Versene solution (1:5 000 concentration)

Protocol

1. Chop the CVS in a sterile plastic Petri dish into very small pieces using a scalpel blade.

2. Add about 2 ml of culture medium to the chopped tissue.

3. Divide between two 5 ml sterile culture flasks and incubate; the small amount of medium allows the tissue to settle down and adhere to the culture vessel surface more quickly.

4. Top up the flasks the following day with 4 ml of medium.

5. Check for cell growth after about 6–7 days and replace the culture medium regularly until the cells are about 75% confluent.

6. Remove the medium from the culture vessel and wash in versene solution.

7. Remove the versene solution

8. Add the trypsin solution and incubate.

9. After a couple of minutes, check to see if the cells are detached from the surface of the culture vessel using an inverted microscope.

10. When most of the cells are floating freely, add culture medium.

11. Divide the cell suspension between about four Petri dishes containing coverslips. Once the cells show adequate growth the cultures are harvested in the same way as for amniotic fluid in situ cultures in *Protocol 2.3*.

Protocol 2.9

Setting up bone marrow and leukemic blood

Equipment

Airflow class II safety cabinet

Centrifuge set at 200 **g** with swinging bucket rotor

Culture vessels, sterile flat-bottomed tubes

Incubator set at 37°C

Pasteur pipettes, plastic, 1 ml

Solutions

Colcemid working solution, 10 µg ml^{-1}

Complete culture medium made up with:

- basal media such as McCoys 5A, 100 ml
- fetal calf serum, 20 ml
- L-glutamine 200 mM, 1 ml
- penicillin solution 10 000 IU ml^{-1}, 1 ml
- streptomycin solution 10 000 µg ml^{-1}, 1 ml

Method

1. Centrifuge the sample at 200 **g** for 10 min and remove the transport medium from the pelleted bone marrow.

2. Add 5 ml of complete culture medium to the number of culture vessels required.

3. Seed with an appropriate amount of bone marrow to give a final concentration of 10^6 cells ml^{-1}. For direct cultures, add 0.1 ml of colcemid immediately to give a final concentration of 0.02 µg ml^{-1} and incubate at 37°C for 1 h. For short-term cultures, incubate at 37°C overnight. The following morning add colcemid as above and incubate for 1–3 h. For overnight with colcemid cultures, add colcemid as above as late in the afternoon as possible, and incubate at 37°C overnight.

4. Harvest cultures using *Protocol 2.11*.

Hints and tips

Bone marrow aspirates and blood samples from patients with leukemia or lymphoma are prone to failure, often due to very low blood cell counts, in particular in patients with myelodysplastic syndromes. Low counts are common following chemotherapy.

It is not uncommon for direct cultures to yield no metaphases, and the benefits of these types of cultures (i.e. speed) are usually outweighed by the very fact that insufficient metaphases are available for analysis, if any at all.

Protocol 2.10

Synchronization of bone marrow cultures using a BrdU/FdU cocktail mixture

Equipment

As in *Protocol 2.9*

Solutions

As in *Protocol 2.9* and:

- FdU working solution (final concentration 4×10^{-3} M), 10 mg in 10 ml distilled water (solution A)

- Uridine working solution (final concentration 3×10^{-3} M), 10 mg in 10 ml distilled water (solution B)

- BrdU working solution (final concentration 10^{-1} M), 30 mg plus 0.1 ml solution A, 2 ml of solution B and make up to 10 ml with distilled water

- Thymidine working solution (final concentration 10^{-3} M), 2.5 mg in 10 ml of distilled water

Protocol

1. Set up a bone marrow culture as described in *Protocol 2.9*, steps 1–3.

2. Add 0.1 ml of the above FdU/BrdU/uridine mixture.

3. Incubate the cultures at 37°C for 14–17 h.

4. Add 0.1 ml of thymidine.

5. Incubate the culture tube at 37°C for 5–7 h.

6. Add 0.1 ml of colcemid to give a final concentration of 0.02 µg ml^{-1} for the final 15 min of culture.

7. Harvest using *Protocol 2.11*.

Hints and tips

BrdU is light-sensitive and some laboratories wrap their cultures in aluminum foil to protect them from the light when incubating with synchronization blocking agents. We have found in this laboratory that it is usually already dark inside an incubator!

Protocol 2.11

Harvest of bone marrow and blood cultures

Equipment

Airflow class II safety cabinet

Centrifuge set at 200 *g* with swinging bucket rotor

Freezer set at −20°C

Incubator set at 37°C

Pasteur pipettes, plastic, I ml

Solutions

Potassium chloride (KCl), 0.075 M, 5.5 g in 1 l of distilled water

Fixative made up of three parts of Analar-grade methanol to one part of Analar-grade glacial acetic acid

Protocol

1. Make up the KCl and ensure it is pre-warmed in the incubator.

2. Make up the fixative and place in the freezer.

3. Following colcemid treatment of the cultures, centrifuge the culture tubes at 200 *g* for 10 min.

4. Remove the supernatant using a plastic pipette and resuspend the cell pellet in 10 ml KCl solution for 10 min at 37°C (blood) or 15 min (bone marrow).

5. Following incubation, add a few drops of chilled fixative, mixing well.

6. Centrifuge the culture tube at 200 *g* for 10 min.

7. Remove the supernatant leaving a small amount just above the pellet. Gently tap the centrifuge tube to resuspend the cell pellet in the remaining supernatant.

8. Carefully add chilled fixative, initially a few drops at a time, with constant agitation to avoid clumps forming. If clumps do start to form, mix the cell suspension thoroughly with a plastic Pasteur pipette. Add further fixative to the tube (up to three-quarters full).

9. Centrifuge the culture tube at 200 *g* for 10 min.

10. Remove the supernatant and replenish with fresh fixative, agitating the tube to obtain an even cell suspension.

11. Repeat steps 9 and 10 a further three times.

12. Prepare slides as in *Protocol 2.12*, or alternatively store the cell suspension at −20°C until slide preparation is required.

Hints and tips

Although the time of KCl treatment may vary according to cell type, cells should not be left in KCl for more than 20 min, since this may be detrimental to chromosome morphology and can also cause cell bursting.

Protocol 2.12

Culture of samples from CLL blood and marrow

Equipment

As for *Protocol 2.9*

Solutions

As for *Protocol 2.9*

PBS

RPMI culture medium:

- RPMI basal medium;

- fetal calf serum, 20%;

- benzylpenicillin (sodium) (BP), 100 000 units l⁻¹;

- streptomycin sulfate, 100 mg l⁻¹;

- L-glutamine 2.3 mg l⁻¹.

TPA stock solution, 100 μg ml⁻¹ dissolved in absolute ethanol or dimethyl-sulfoxide (DMSO); store frozen in the dark

TPA working solution, 2–5 μg ml⁻¹ in RPMI made up prior to use

Protocol

1. Set up blood cultures and/or marrow cultures as in *Protocol 2.9*.

2. Add 0.2 ml TPA working solution to one blood and bone marrow culture, and leave one culture tube with no mitogens.

3. Incubate at 37°C for 3–5 days. It is useful to use a range of times if possible.

4. Add colcemid to the cultures for 3–4 h.

5. Centrifuge the culture tubes at 200 g for 10 min and add PBS, which helps to clean up the preparations.

6. Centrifuge the tubes again.

7. Remove the supernatant and add KCl for 20 min.

8. Continue the harvesting and slide-making procedures as in *Protocol 2.11* (from Step 5) and *2.14*.

Hints and tips

During slide-making, be careful not to overspread the slides as TPA can cause the bursting of cells.

Protocol 2.13

Preparation and culture of samples from lymphomas

Equipment

As for *Protocol 2.12*

Culture dish

Scalpel blades

Solutions

As for *Protocol 2.12*

Protocol

1. Retain the transport medium with the node tissue, as this often contains a number of cells that can be used in culture.

2. Place the biopsy into the culture dish, with 1–2 ml RPMI culture medium and remove the fat, blood and connective tissue. Begin to chop up the tissue with a scalpel blade or sharp scissors.

3. Chop up the tissue thoroughly using a scalpel blade. There should be a bursting of cells from the tissue into the small amount of medium.

4. Collect up the cells by pushing into the crevice of the Petri dish.

5. Add the transport medium to these cells and place into a sterile centrifuge tube.

6. Centrifuge at 200 g for 10 min, and resuspend the cells in fresh RPMI.

7. Seed some culture vessels equally with the cell/medium mixture to make an optimum concentration of 10^6 cells ml^{-1} (as for bone marrow in *Protocol 2.9*).

8. To one tube add 0.2 ml of TPA, leave one unstimulated and to another add 0.1 ml of colcemid immediately, and this latter tube should be incubated overnight, harvested as in *Protocol 2.11* and slides made as per *Protocol 2.14*.

9. Incubate the other tubes for 3–5 days as in *Protocol 2.12*.

10. Continue to harvest as in *Protocol 2.12*.

Hints and tips

If the lymph node is hard and not easy to chop up, it is likely that the cells will not grow. It is possible to use collagenase to disaggregate the cells, but failure still often occurs even following this treatment.

Protocol 2.14

Slide-making

Equipment

Centrifuge set at 200 *g* with a swing out rotor

Glass microscope slides

Lint-free cloth

Pasteur pipettes (plastic)

Phase contrast microscope

Refrigerator at 4°C

Solutions

Distilled water

Fixative made up of three parts Analar-grade methanol to one part Analar-grade glacial acetic acid

Methanol

Protocol

1. Thoroughly clean the slides using methanol and wipe clean with a lint-free cloth to ensure the slides are grease-free.

2. Place the slides in a rack and put this in a glass Petri dish full of distilled water in order to cover the slides completely.

3. Place this in a refrigerator and cool before use.

4. Alternatively, slides can be used dry depending on the tissue being spread. Bone marrow and blood from leukemic patients require cold wet slides for optimum spreading. Lymphocytes, amniocytes and CVS cultures can be spread onto dry clean slides. A water bath can be employed to provide a humid atmosphere during spreading if required.

5. If the cell suspension has been stored at −20°C, apply steps 6 and 7. If the cell suspension has been freshly prepared, proceed directly to step 8.

6. Centrifuge the culture tube at 200 *g* for 10 min, remove the supernatant and replenish with fresh fixative.

7. Centrifuge the culture tube at 200 *g* for 10 min and remove the supernatant to just above the cell pellet.

8. Add a few drops of fixative to the pellet (approximately 0.5 ml depending on the size of the pellet).

9. Hold a slide horizontally with forceps and drop the cell suspension at each end of the slide, to give two drops, from a Pasteur pipette held just above the slide.

10. Label the slide. Wipe the excess water from the back of it and immediately place upon a hotplate at approximately 70°C.

11. Either leave the slides to dry on the hotplate overnight, or put into a slide rack and place in an oven at 60–70°C overnight.

Hints and tips

If the chromosomes have not spread adequately, the fixative evaporation time should be increased, which can be rectified by various methods, some of which are shown below:

■ Add extra fixative onto the slide after the drop of suspension, immediately prior to the drop drying out.

■ Increase the humidity by making the slides over a waterbath, or some laboratories even have a specially fitted spreading room where temperature and humidity are controlled.

■ Keep the slide horizontal rather than at an angle while the suspension dries out (i.e. leave it flat on the bench).

If chromosomes are overspread or broken, the evaporation time of the fixative should be decreased, which can be done in the following ways:

■ Drop the suspension onto slides and immediately place the slides on a warm surface, such as a hotplate.

■ Allow the dropped suspension to run down an angled slide (i.e. hold with forceps at about 45°).

If the cells and metaphases appear too crowded, add extra fixative to the suspension. If the suspension is too sparse, centrifuge the tube again and add less fixative to the cell pellet than previously.

Chapter 3

Staining and banding of chromosome slides

Barbara Czepulkowski

1. Introduction

All banding techniques can be applied to most tissues, including amniotic fluid cultures, chorionic villi, solid tissues, peripheral blood, bone marrow cultures, lymph node preparations and leukemic blood chromosomes. Various adaptations need to be implemented between each different tissue type, although the principles remain the same for all banding techniques.

A number of treatments which involve enzymatic digestion and/or denaturation, followed by the use of a DNA-specific dye, give unique transverse light and dark staining bands of varying intensity (Craig and Bickmore, 1993). Banding reflects the varying degrees to which the chromatid of the chromosome differs in base composition, replication time, chromatin conformation and gene density.

G-bands are known to be late replicating and contain relatively highly condensed heterochromatin, while the pale G-bands (or dark-staining R-bands) replicate early in S phase, and possess fewer condensed chromatin structures. The late-replicating, highly condensed DNA is transcriptionally less active, and genes are known to concentrate in the pale G-band areas.

G- and R-bands were thought to be AT and CG rich, respectively, as the dye quinacrine used to produce the Q-banding pattern preferentially binds to AT-rich DNA, while R-banding can be initiated using chromomycin, which preferentially binds to CG-rich DNA. However, it is now understood that G/Q-band DNA is only slightly richer in AT concentration than R-band DNA (Saitoh and Laemmli, 1994). Chromatin loops are attached to the chromosome scaffold at special scaffold attachment regions (SARs). The DNA which forms G/Q-bands has more closely spaced SARs, giving tighter loops (Q loops). Those regions with more unfolded SARs are associated with R-bands. There are more SARs per unit length in G-bands than R-bands (see *Figure 3.1*).

Chromosomes prepared from malignant cells are particularly sensitive to various treatments involved in G-banding and chromosome morphology can often be fuzzy with indistinct bands. *Protocol 3.3* provides a G-banding technique which has been modified for use with malignant chromosomes, using Leishman's stain (*Protocol 3.2*), which appears to produce sharper banding than Giemsa, although this is generally a matter of personal choice.

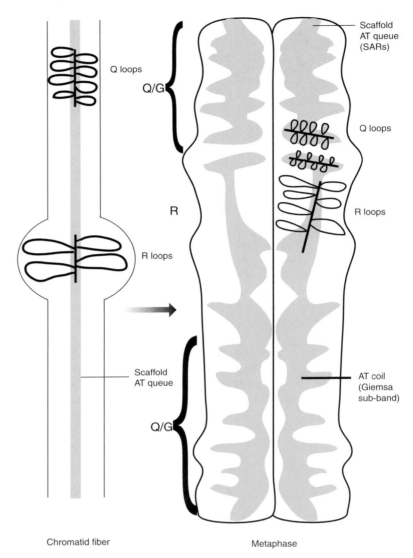

Chromatid fiber   Metaphase

Figure 3.1

Structural basis of chromosome bands. Chromatin loops are attached to the chromatid scaffold at SARs. The G/Q-bands are formed from closely spaced SARs, with tight loops. There are consequently more SARs per unit length in G/Q-bands than in R-bands. (Figure adapted from Strachan, T. and Read, A.P. (1996) Human Molecular Genetics. BIOS Scientific Publishers, Oxford.)

2.   Banding techniques and their applications

This chapter will provide protocols for the most frequently used techniques applied in clinical situations, as shown below.

- Solid staining – Although apparently outdated since the inception of banding, this still can be applied to chromosome breakage syndromes and fragile site studies, as banding sometimes masks small chromatid gaps and breaks in the chromosomes.
- G-banding – the most widely used technique for the study and recognition of mammalian chromosomes.
- Q-banding – this is a fluorescent staining technique using quinacrine, which produces a similar banding pattern to G-banding. It is useful in the study of polymorphic variants on chromosomes, regions such as those on chromosomes 1, 3, 9, 16, all the acrocentrics and the Y chromosome, which usually fluoresce very brightly. Quinacrine binds to DNA by intercalation of external ionic banding.
- C-banding – this produces dark staining of the constitutive heterochromatin located at the centromeres of all the chromosomes. Again, as with Q-banding, polymorphic variants can be studied with this technique. The chromosomal DNA is preferentially denatured in alkali and lost from the non-C-band regions. The reasons for the reaction are unclear but may involve histone proteins closely bound to the C-band heterochromatin.
- R-banding – this produces a pattern which is a reverse pattern of that observed with G-banding. A few centers prefer to use this method rather than G-banding as their routine identification banding. The advantage of this technique is that the telomeric regions of several chromosomes, which stain faintly with G-banding, are stained darkly.
- NOR staining – the NOR of chromosomes (see *Chapter 1*) are known to contain genes for *18S* and *28SRNA*. The active transcription sites of these chromosomes can be stained selectively using silver nitrate. This technique is crucial in identifying small bi-satellited marker chromosomes. The staining pattern is seen as small dot-like structures on the stalks of acrocentric chromosomes 13, 14, 15, 21 and 22. As with C- and Q-banding, polymorphisms can be studied with this technique, as individual staining patterns are apparent in different samples.
- Distamycin A-4,6-diamino-2-phenyl-indole (DA-DAPI) banding – this is a technique most useful for identifying small marker chromosomes derived from chromosome 15, but also stains the heterochromatic regions of chromosomes. Both DA and DAPI have an affinity for AT base pairs, at similar but not identical sites.

For further details of other staining techniques which are beyond the scope of this volume, the reader is referred to Verma and Babu (1989) and Benn and Tantravah (2000).

Acknowledgments

I am indebted to Angela Douglas of the Liverpool Women's Hospital who kindly provided some of the excellent photographs for this chapter. For these I am extremely grateful.

References

Benn, P.A. and Tantravahi, U. (2000) Chromosome staining and banding techniques. In: *Human Cytogenetics 3e: Constitutional Analysis* (Vol. 1), Practical Approach series no. 240. Oxford University Press, Oxford.

Craig, J.M. and Bickmore, W.A. (1993) *Bioessays,* **15**: 349–354.

Saitoh, Y. and Laemmli, U. K. (1994) *Cell* **76**: 609–622.

Verma, R.S. and Babu, A. (1989) *Human Chromosomes: Manual of Basic Techniques.* Pergamon Press, New York.

Protocol 3.1

Solid staining

Equipment

Coplin jars

Coverslips

Fixative made up of three parts Analar-grade methanol and one part Analar-grade glacial acetic acid

Hotplate at 60°C

Mountant (DPX or XAM)

Wash bottles (for buffer wash if required)

Xylene (if required for removing coverslips)

Solutions

Distilled or de-ionized water

Giemsa stain – 5 ml Giemsa plus 45 ml of buffer (as above)

Phosphate buffer (pH 6.8) – 0.025% M KH_2PO_4 (3.4 g l^{-1}) titrated to correct pH using 50% NaOH

Protocol

1. Place slides in the Giemsa solution in a Coplin jar for 8 min.

2. Rinse the slides twice using the phosphate buffer (either in two separate Coplin jars filled with buffer or by using plastic wash bottles).

3. Air-dry by wiping the back of the slide and placing on a hotplate.

4. Mount the slides with a coverslip using DPX mountant (see *Figure 3.2*).

Hints and tips

If further treatments are required, do not mount with DPX but use distilled water, place the coverslip on top, examine the slide, then air dry once more and the slide can be de-stained using fixative.

Occasionally it happens that the slide has already been mounted with DPX mountant, and it is still necessary to de-stain the slide. The coverslip can be removed by soaking in a Coplin jar of xylene (this may take some time).

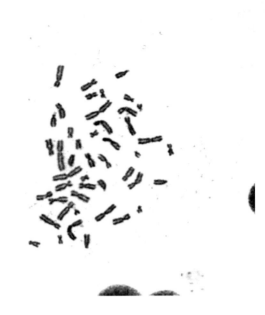

Figure 3.2

A metaphase spread demonstrating solid staining of chromosomes.

Protocol 3.2

Preparation of Leishman's stain for G-banding

Equipment

Conical flask (1–2 l)

Filter paper (30 cm diameter)

Filter funnel (large)

Hotplate at 70°C

Magnetic stirrer (optional)

Parafilm or aluminum foil

Universal tubes (plastic, 30 ml)

Solutions

Absolute methanol (Analar grade)

Leishman's stain (powder)

Protocol

1. Add 1.5 g Leishman's stain to 500 ml Analar-grade absolute methanol in a conical flask. This should be done gradually, while agitating the flask periodically, to ensure that the stain is completely dissolved.

2. Cover the neck of the conical flask with Parafilm or aluminum foil and leave stirring (optional) on a hotplate overnight at 70°C.

3. Filter the solution through one or two layers of filter paper into sterile Universal tubes.

4. Store, preferably in a dark cupboard.

Hints and tips

Small aliquots are beneficial as the staining power of Leishman's decreases with exposure to air.

It is also an advantage to fill the Universal tube up as close to the top as possible (without flooding your laboratory with stain, as has been known to happen!).

Protocol 3.3

G-banding

Equipment

Coplin jars (prepare the following Coplin jars in the order given):

■ 1–2 ml trypsin solution in 50 ml saline solution

■ 50 ml saline solution (for rinsing)

■ 50 ml buffer pH 6.8 (for rinsing) × 2

■ 50 ml distilled water (for rinsing)

■ 50 ml buffer pH 6.8 (for diluting Leishman's stain)

Hotplate at 70°C

Slide rack for horizontal staining

Wash bottles

Solutions

Absolute methanol (Analar grade)

Buffer pH 6.8 prepared from Gurr's buffer tablets

Distilled water

Glacial acetic acid (Analar grade)

Leishman's stain (from *Protocol 3.2*). Use freshly prepared diluted 1:5 with buffer pH 6.8 (or Giemsa, 2.5 ml in 45 ml pH 7.0 buffer)

Mountant (DPX or XAM)

Saline solution (8.5 g of NaCl in 1 l of distilled water)

Trypsin solution (reconstitute a vial of 'Bacto-trypsin' with 10 ml de-ionized or distilled water).

Protocol

1. Stain the slide horizontally with buffered Leishman's stain for 1 min.

2. Dry the slide on a hotplate and then de-stain by flushing first with glacial acetic acid, and then with methanol. Flush over a sink and keep the tap water running to drain away excess acetic acid and methanol.

3. Dry the slide on the hotplate then dip into the Coplin jar of trypsin for 20–50 s (depending on the preparation).

4. Rinse the slide in the Coplin jar containing saline.

5. Rinse the slide in the Coplin jar containing buffer pH 6.8.

6. Stain the slide horizontally for 1–1.5 min with buffered Leishman's (or for 5 min with buffered Giemsa).

7. Rinse the slide in the second jar of buffer pH 6.8.

8. Finally rinse the slide in the distilled water jar.

9. Mount a coverslip on the wet slide to allow examination of banding. This enables further re-banding if necessary.

10. If banding has been successful, remove the coverslip carefully and dry the slide on a hotplate. If further banding is necessary, repeat the procedure from step 3.

11. Mount the slide with a coverslip using DPX or XAM mountant (see *Figure 3.3*).

Hints and tips

Giemsa can be used to replace Leishman's stain, using 2.5 ml of Giemsa and 45 ml of buffer (pH 7.0).

Steps 1–3 are optional and it is possible to band the slide without pre-treating with the solid-staining procedure (a matter of choice). If the slides are more than 2 days old, steps 1–3 can usually be dispensed with, and the slide can be dipped directly into the prepared trypsin solution.

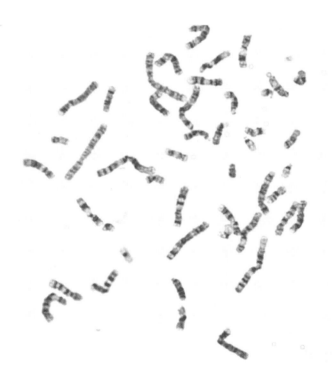

Figure 3.3

A metaphase spread demonstrating G-banding of chromosomes.

The time in trypsin will vary enormously depending upon the tissue being banded. Bone marrow cultures usually require the longest time.

It must be remembered that the banding will not look as crisp and clear under water as it would in mountant, and hence the examination is only a rough guide to the quality.

Trouble-shooting advice

When the chromosomes appear too dark and the bands are indistinct, de-stain the slide, then repeat the procedure from step 3, using about 30 s trypsin time, and only stain for about 30 s.

If the chromosomes appear too pale, just re-stain the slide for about another 15–30 s, checking the slide each time underwater before mounting with DPX.

Always try to be cautious when trying a first batch of slides. Under-band initially if unsure of the optimum trypsin treatment time, which again can vary depending upon the conditions and the sample. When chromosomes appear indistinct around their edges (over-banded), it is probably due to too long an exposure time to trypsin. Under-banded slides can be rescued, over-banded slides cannot!

Chromosomes can sometimes appear too puffy even though the banding pattern is reasonable and this is usually the fault of the buffer being of an incorrect pH. Check the buffer pH or make fresh buffer if necessary.

Protocol 3.4

Q-banding

Equipment

Aluminum foil

Coplin jar

Coverslips (extra-thin size '0')

Filter paper

Solutions

Distilled water

MacIlvaine's buffer (pH 5.6) – 0.1 M anhydrous citric acid (19.2 g l^{-1} – solution A) and 0.4 M anhydrous sodium phosphate dibasic (Na$_2$HPO$_4$ 56.8 g l^{-1} – solution B). Use 92 ml of solution A and 50 ml of solution B, and adjust the pH to 5.6 if necessary.

Quinacrine dihydrochloride solution – 0.5 g quinacrine dihydrochloride in 100 ml of distilled water (store the solution in an aluminum foil-covered container in the refrigerator).

Rubber cement (or nail varnish)

Tap water

Protocol

1. Place the slide in a Coplin jar of the quinacrine solution for 10 min.

2. Rinse the slide in tap water to remove the excess stain.

3. Place the slide in the MacIlvaine's buffer for 1–2 min.

4. Mount the slide with a few drops of the buffer using an extra-thin coverslip.

5. Remove as much of the excess buffer as possible by pressing the slide gently between filter paper.

6. Seal the coverslip with rubber cement or nail varnish.

7. Examine using a fluorescence microscope (see *Figure 3.4*).

Hints and tips

The images observed fade quite quickly so it is advisable to perform photography of the cells being examined.

There is no need to age the slides using this method, and it works equally well with older slides, and even poor preparations show some Q-banding.

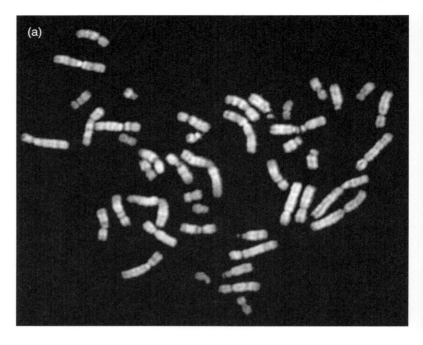

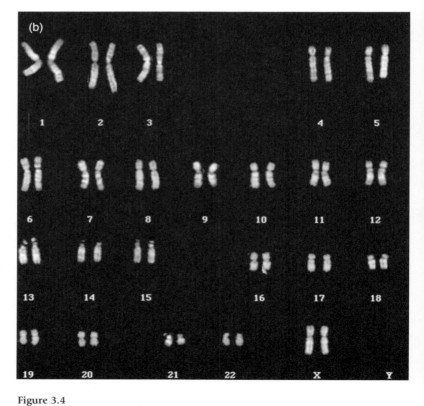

Figure 3.4

(a) A metaphase spread and (b) a karyotype demonstrating Q-banding of chromosomes. (Photograph supplied by Angela Douglas of the Liverpool Women's Hospital.)

Protocol 3.5

C-banding

Equipment

Coplin jars

Coverslips

Waterbath at 50°C and then 60°C

Solutions

Alcohol series, 70% and absolute alcohol

Barium hydroxide solution ($BaOH_2$ 5%) – 5 g barium hydroxide in 100 ml distilled water

Distilled water

Giemsa solution – 2.5 ml Giemsa, 45 ml Sorensen's buffer

Hydrochloric acid (HCl) 0.2M – 10 ml 2 M HCl in 100 ml distilled water

Sorensen's buffer 0.06 M, pH 6.5 made up as below:

- 5.6 g potassium phosphase monobasic (KH_2PO_4)
- 2.64 g sodium phosphate dibasic (Na_2HPO_4)
- 1 l distilled water

$2 \times$ SSC made up as below:

- 17.5 g sodium chloride
- 8.8 g sodium citrate
- 1 l distilled water

XAM or DPX mountant

Protocol

1. Place the slides in a Coplin jar with 0.2 M HCl for 1 h at room temperature.

2. Rinse the slides thoroughly with distilled water and allow to air dry.

3. Incubate the slides in $BaOH_2$ solution for 5–15 min at 50°C in a waterbath. (Alternatively incubate at room temperature for 30–40 min.)

4. Rinse the slides a number of times with distilled water.

5. Pass the slides through the 70% and absolute alcohol series, then allow to air dry.

6. Incubate the slides in $2 \times$ SSC at 60°C in a water bath for 2 h.

7. Rinse the slides in distilled water and air dry.

8. Stain the slides with 5% Giemsa solution for 10–15 min.

9. Mount the slides with DPX mountant (see *Figure 3.5*).

Hints and tips

Always prepare the $BaOH_2$ freshly prior to use as it interacts with CO_2 in the air and forms barium carbonate, which is only sparingly soluble in water.

During use, the container with $BaOH_2$ should be closed to prevent the deposits of barium carbonate crystals forming on the slides. These should be removed by the washes above, but if they persist, a brief rinse in a dilute HCl solution will remove them.

Overnight incubation in $2 \times SSC$ gives better differentiation of C-bands but more background stain also appears.

Use slides that are at least 7 days old for the best results.

Incubation at room temperature of $BaOH_2$ preserves chromosome morphology.

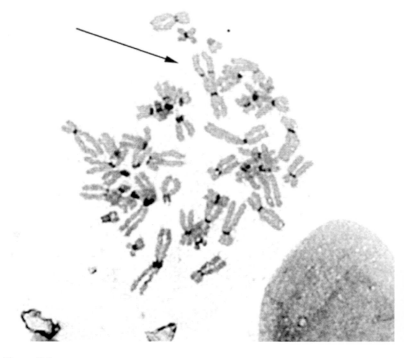

Figure 3.5

A metaphase spread demonstrating C-banding of chromosomes. The arrow indicates an abnormal dicentric chromosome 1. (Photograph supplied by Angela Douglas of the Liverpool Women's Hospital.)

Protocol 3.6

R-banding

Equipment

Coplin jars

Coverslips

Waterbath at 85°C

Solutions

Distilled water

Sorensen's buffer 0.06 M, pH 6.5 made up as below:

- ■   5.6 g potassium phosphase monobasic (KH_2PO_4)
- ■   2.64 g sodium phosphate dibasic (Na_2HPO_4)
- ■   1 l distilled water

Giemsa solution (make fresh) – 2.5 ml Giemsa and 45 ml buffer pH 6.8

XAM or DPX mountant

Protocol

1. Age the slides for at least a week (fresh slides are not suitable for the hot waterbath treatment).
2. Heat the Sorensen's buffer in a Coplin jar to 85°C using the waterbath.
3. Incubate the slides in heated buffer for 8 min.
4. Rinse the slides with Sorensen's buffer.
5. Stain the slides in Giemsa solution for 6–10 min.
6. Rinse the slides in distilled water
7. Mount the slides using DPX or XAM mountant (see *Figure 3.6*).

Hints and tips

The incubation time in Sorensen's buffer at 85°C is crucial, and does depend on the age of the slide. If slides are older than 7–10 days, the incubation period should be reduced.

(a)

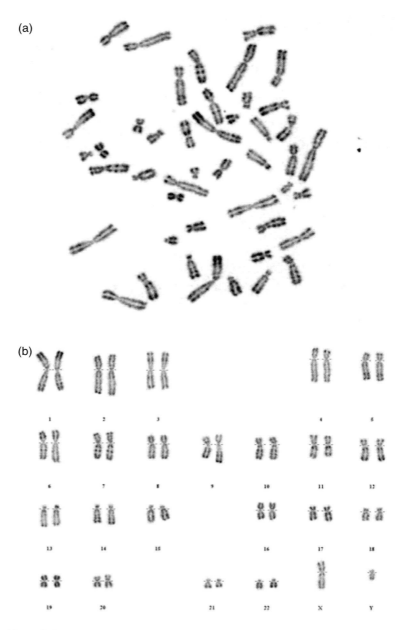

(b)

Figure 3.6

(a) A metaphase spread and (b) a karyotype demonstrating R-banding of chromosomes. (Photograph supplied by Angela Douglas of the Liverpool Women's Hospital.)

Protocol 3.7

Nucleolar organizer region (NOR) staining

Equipment

Coverslips

Oven at 60°C (or hotplate at 60°C)

Solutions

Ammoniacal silver – 4 g silver nitrate in 5 ml of concentrated ammonium hydroxide. Slowly add 7.5 ml of distilled water (there should be no precipitate). Store the solution in a dark bottle in the refrigerator

Distilled water (in squeezy bottle)

DPX or XAM mountant

Formalin 3% solution (pH 4.5) – 3 ml formalin in 97 ml distilled water

Silver nitrate solution (50%) – 5 g of silver nitrate in 10 ml of distilled water

Protocol

1. Place three drops of the 50% silver nitrate solution on the slide.

2. Place a coverslip gently on top of this solution and place the slide in an oven at 60°C (or hotplate at 60°C) until the silver nitrate becomes crystalline.

3. Remove the coverslip by rinsing the slide with distilled water using a squeezy bottle.

4. Place three drops of the 3% formalin solution onto the slide and also three drops of ammoniacal silver solution. Place a coverslip gently on top of these solutions on the slide.

5. Using a microscope (low power ×10) the staining reaction can be observed. During a 30–60 s period the cells begin to develop a golden brown color. When this has been achieved, rinse the slide well in distilled water to remove the coverslip and excess solutions.

6. Air-dry the slide and mount using DPX or XAM mountant (see *Figure 3.7*).

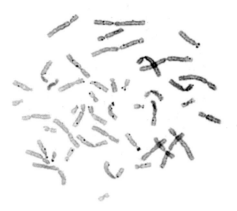

Figure 3.7

A metaphase spread demonstrating NOR staining of chromosomes. (Photograph supplied by Angela Douglas of the Liverpool Women's Hospital.)

Protocol 3.8

DAPI/distamycin A staining

Equipment

Coplin jars

Coverslips

Fluorescence microscope

Staining rack

Squeezy bottle with MacIlvaine's buffer (pH 7.5)

Solutions

DAPI stock solution – 2 mg in 10 ml distilled water (add a small amount of methanol which aids the dissolving of DAPI). Store this stock solution in the freezer. For the working solution, use a 1 in 1000 dilution in MacIlvaine's buffer.

Distamycin A solution – 2 mg of distamycin A in 10 ml MacIlvaine's buffer (pH 7.5). This stock solution should be kept in the freezer. For the working solution use a 1 in 10 dilution of the stock in MacIlvaine's buffer.

Glycerol.

MacIlvaine's buffer (pH 7.5) – 0.1 M anhydrous citric acid (19.2 g l^{-1} – solution A) and 0.2 M anhydrous sodium phosphate dibasic (Na$_2$HPO$_4$ 28.4 g l^{-1} – solution B). Use 80 ml of solution A and 920 ml of solution B, and adjust the pH to 7.5 if necessary.

Protocol

1. Soak the slide in MacIlvaine's buffer for 10 min in a Coplin jar.
2. Stain the slide for 10 min with distamycin A using a staining rack.
3. Rinse the slide in MacIlvaine's buffer using a squeezy bottle.
4. Stain the slide for 10 min with the DAPI solution using a staining rack.
5. Rinse the slide with MacIlvaine's buffer using a squeezy bottle.
6. Mount the slide using a 1:1 glycerol:MacIlvaine's buffer solution.
7. Use a fluorescence microscope to view the slide.

Hints and tips

If the chromosomes show a Q-banding appearance, stain for longer with distamycin A.

If the chromosomes are too dull, reduce the distamycin A staining time.

Chapter 4

Types of abnormalities observed in chromosomes

Barbara Czepulkowski

1. Introduction

Mutations affecting the genomic DNA can result in the loss or gain of a whole chromosome, or of smaller segments, including duplications. Rearrangements which do not cause a loss of material can also be observed, for instance in translocations, insertions and inversions (*Figure 4.1*). If the abnormality is large enough to be visible using the light microscope it is termed a chromosome abnormality or aberration. The resolution of the light microscope allows changes of 4 Mb or larger to be observed. However, if the changes are smaller, that is they occur at a molecular level, then other techniques are employed to visualize them – molecular techniques such as polymerase chain reaction (PCR) and/or single locus probes using fluorescent *in situ* hybridization (FISH). This chapter will also cover the nomenclature used for normal and abnormal chromosomes.

2. Types of chromosomal changes

Chromosomal changes can be present in cells throughout the body (i.e. an individual would be born with the change) and this would be termed a constitutional abnormality. Alternatively, the change can arise in a subset of cells or tissues, and these changes are known as acquired abnormalities. The constitutional change would be present very early in development, probably from an abnormality present in a sperm or egg cell, abnormal fertilization or an abnormal event in the early embryo, whereas acquired abnormalities can occur at any time depending on the disease type.

Normal chromosome constitutions in mammalian chromosomes are described by a karyotype where the number of chromosomes is stated, which in humans is 46 and then the sex chromosomes XX or XY. If there is an abnormality, this is also described in the karyotype following guidelines laid down by the International System for Chromosome Nomenclature (ISCN, 1995), as a result of several international conferences. Abnormalities fall into two main categories, numerical and structural. This chapter is by no means an exhaustive guide to the many different ways that chromosomes can be altered, and is not intended as a replacement for the ISCN, but as an introductory guide. Further information and greater detail can be found in the ISCN, the definitive guide to terminology and writing of the karyotype.

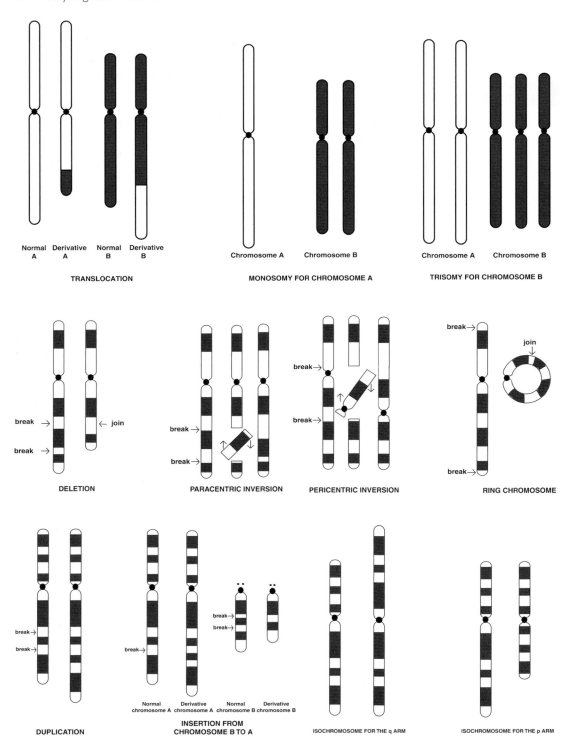

Figure 4.1

Types of chromosomal changes.

2.1 Numerical chromosomal changes

Numerical changes involve a change in chromosome number without actual breakage of the chromosome. There are three different classes of numerical changes: polyploidy, aneuploidy and mixoploidy.

Polyploidy

The most commonly observed polyploidy is triplody ($3n$, where n is the haploid number; in the case of human chromosomes $n = 23$) where two sperm fertilize the same egg, or alternatively by fertilization with an abnormal diploid gamete. Tetraploidy ($4n$) is usually a result of failure to achieve the first zygotic division; the DNA has replicated, but cell division does not subsequently take place. Constitutional polyploidy is rare but can exist in regenerating cells, such as those of the liver, and megakaryocytes which arise in the bone marrow with extremely large nuclei, giving rise to numerous platelet cells which themselves lack a nucleus and are termed nulliploid. Polyploidy in an individual would not be compatible with survival. If chromosome analysis is carried out on aborted fetuses, the karyotype can show polyploidy, usually tetraploidy.

Aneuploidy

Aneuploidy is the result of extra copies of a single chromosome being present (e.g. an extra copy in addition to the two normal homologs is termed trisomy, two extra copies would be termed tetrasomy and so on), or alternatively the complete loss of a homolog, termed monosomy. If both homologs are missing this is called nullisomy. Cells arising from malignant conditions such as acute leukemia often display severe aneuploidy with multiple chromosome abnormalities depending on the stage of the disease.

This type of cell arises from either:

(i) nondisjunction, where the paired chromosomes do not separate at meiosis I, or the paired sister chromatids fail to separate at meiosis II or mitosis; or

(ii) anaphase lag, where the chromosome is not incorporated into one of the daughter nuclei following cell division. This occurs because of a delay in the movement (lagging) of the chromosome during anaphase. The chromosome is then lost.

Mixoploidy

Mixoplody occurs because of mosaicism, a term applied to an individual who possesses two or more genetically different cell lines from a single zygote, or occasionally because of chimerism, where two or more differing cell lines have resulted from different zygotes.

Aneuploid mosaics are quite common and can be attributed to nondisjunction or chromosome lagging in the mitotic divisions of early embryonic cells, giving rise to trisomic cells and monosomic cells (the latter usually being lost after a short interval). If an individual has a small proportion of trisomic cells, for instance trisomy 21, Down syndrome, the congenital problems and phenotypic effect will not be as severe as in an individual whose whole constitutional karyotype was made up of trisomy 21 in every cell.

2.2   Structural chromosomal changes

Structural chromosomal abnormalities arise from chromosomal breakage. There are a number of ways in which a chromosome can be altered following breakage (see *Figure 4.1*). If a break occurs at one point in a chromosome, the breakpoint is normally repaired rapidly by the repair enzymes described in *Chapter 1, Section 5*. However, occasionally terminal deletions occur, where a distal fragment (the portion furthest from the centromere) breaks and, as it then has no centromere, is subsequently lost. This is known as a terminal deletion, and although the remaining portion of chromosomal material has a centromere, the lack of a functional telomere makes the chromosome unstable and in these cases the chromosome is often degraded.

If breakage of a chromosome occurs at different points (i.e. on two different chromosomes), the repair enzymes have difficulty in performing the 'jigsaw puzzle' and often join the wrong ends together, resulting in a number of structural abnormalities. These structural changes may be balanced when there is no net gain or loss of chromosomal material, or unbalanced when there is.

Two breaks in a single chromosome

Two breaks occurring in a single chromosome can result in three types of structural abnormality:

(i)   Inversion – the chromosomal segment which has broken in two places is flipped upside down (inverted) and then rejoined, resulting in a chromosome which is the same shape and size as its homologs, but the genetic material is now in the wrong order on the chromosome. This can cause problems at meiosis (see *Chapter 1, Section 6.2*). If an inversion occurs in only one of the chromosomal arms, either the p or q arm, it is termed a paracentric inversion. If the centromere is involved in the inversion (i.e. the breaks occur in both the p and q arm, on either side of the centromere), this is termed a pericentric inversion. (The author was never able to remember which way round this was, but using the fact that pericentric contained an 'e' as the second letter of the word and so did centromere, it all suddenly became clear!)

(ii) Interstitial deletion – two breaks occur, and two of the resulting fragments repair without the presence of the intervening fragment. The only possible interstitial deletions arise from breaks in the same arm of a chromosome, as a centromere is required for chromosomes to function. The excluded portion lacks a centromere (known as an acentric fragment) and would be lost in the next cell division.

(iii) Ring chromosome – as the name suggests, this is the result of breakage occurring either side of the centromere and the two ends of the segment being repaired to form a circular chromosomal fragment, again only with a centromere would the abnormal chromosome be able to pass through subsequent cell divisions.

Two or more breaks in different chromosomes

When breaks occur in two different chromosomes, abnormal chromosomes are formed containing material from other chromosomes, this is termed a translocation. Three types of translocation are described below.

Reciprocal This is a balanced translocation where material (the broken acentric fragments) distal to the two breakpoints is exchanged. Breaks can occur in both the long and short arms of the chromosomes. Although carriers of balanced reciprocal translocations can be asymptomatic, some cases can involve a disruption of the DNA sequences at the relative breakpoints, which can lead to inactivation of gene expression or inappropriate gene expression. However, when the carrier is asymptomatic, offspring born to such individuals may be at risk, as during the formation of eggs or sperm unbalanced gametes may result (see *Figure 4.2*) which if fertilized by a normal gamete will give rise to partial trisomy and partial monosomy. If the chromosomal segments are large, the unbalanced fertilized products would be unviable. In only a very rare instance could a very small exchange of chromosomal material at terminal regions of chromosomes give rise to a fertilized zygote that reaches term. The usual scenario is that the carrier and partner experience spontaneous abortions, and examination of both partners often reveals a balanced translocation which would provide the underlying reason for their fertility problems. The likelihood of producing a normal fetus with either a normal karyotype or with the balanced translocation is 50:50.

Centric (Robertsonian) This type of rearrangement occurs when the breaks occur near the centromere of two acrocentric chromosomes, and the large fragments of two chromosomes fuse together. The breaks are usually just above or in the centromere and the resulting abnormal chromosome may be dicentric (have two centromeres). In each instance the acentric fragments are lost. Although there is loss of chromosomal material, Robertsonian translocations are considered balanced, as they generally produce no phenotypic effects. As was the case in the balanced reciprocal translocation, an asymptomatic carrier produces unbalanced gametes, which may in turn result in fertility problems (see *Figure 4.3*).

Insertion This is the result of three breaks, two in one chromosome. The resulting DNA segment can then be inserted into a break occurring in a second chromosome. Again, the offspring may be at risk of partial monosomy and/or partial trisomy.

2.3 Uniparental diploidy, disomy and genomic imprinting

Very occasionally, although cytogenetically an individual appears to have a normal karyotype, either 46,XX, or 46,XY, the karyotype is in fact abnormal due to an unequal contribution from either parent, the extreme case being uniparental diploidy, where all the chromosomes are derived from one parent. This results in failure of embryonic development in humans. The more common instance is uniparental disomy where two copies of a specific chromosome are inherited from one parent or uniparental isodisomy where two identical copies of a single homolog arise from either parent. This often contributes to disease.

For the great majority of mammalian genes, the expression of most alleles does not depend upon whether they have been derived maternally or paternally. A few mammalian autosomes, however, differ in that expression

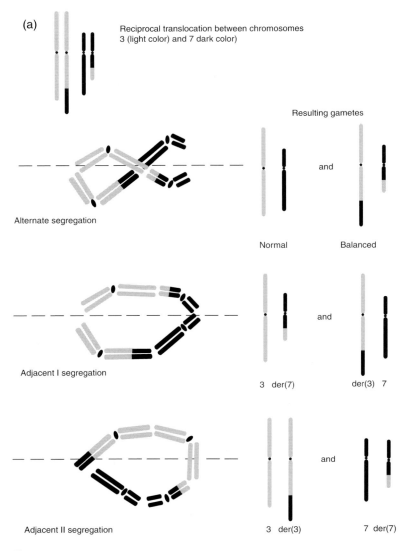

(a)

Reciprocal translocation between chromosomes 3 (light color) and 7 dark color)

Resulting gametes

Alternate segregation

Normal   Balanced

Adjacent I segregation

3   der(7)   der(3)   7

Adjacent II segregation

3   der(3)   7   der(7)

Figure 4.2

(a) The behavior of chromosomes involved in a translocation t(3;7) at segregation during meiosis, and the resulting gametes. (Figure adapted from Connor and Ferguson-Smith (eds). Essential Medical Genetics, 5th edn. 1997, Blackwell Science, Oxford.)

does depend on whether inheritance was from the mother or father as the maternal allele alone is expressed in certain cells and for other cells it is the paternal allele that is expressed. This phenomenon is called genetic imprinting. Imprinting of genetic material occurs during gametogenesis and results in the silencing of transcription along the imprinted region. In these imprinted regions, therefore, the genetic information is inactivated when inherited from one of the parents, but not from the other, and only one copy of the gene is transcribed. Because of this, maternally and paternally derived genetic material is required for normal development of a conceptus; therefore, where a part, or all, of a homologous chromosome pair that is

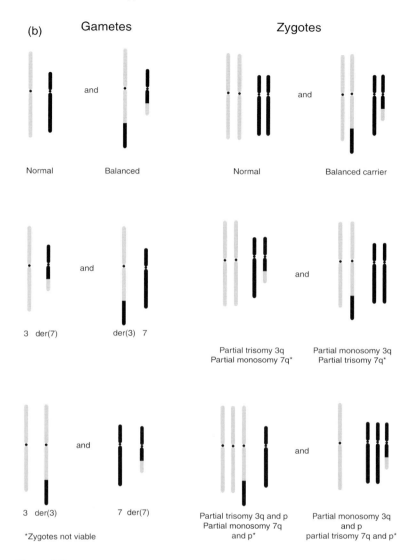

Figure 4.2

(b) When fertilized by normal gametes, the resulting zygotes are also shown. (Figure adapted from Connor, M. and Ferguson-Smith M. (eds) (1997) Essential Medical Genetics, *5th Edn. Blackwell Science, Oxford).*

subject to genomic imprinting is inherited from only one parent, known as uniparental disomy, abnormality can result.

The most extreme example of this is uniparental disomy for the full diploid complement. Paternal uniparental disomy, where all 46 chromosomes are contributed by the father, results in a hydatiform mole, an abnormal pregnancy where the placenta develops abnormally and the chorionic villi become very fluid filled (hydropic), the trophoblast is hyperplastic and there is no evidence of recognizable fetal parts. The incidence of complete mole is about 1:2 000 pregnancies and is thought to occur when

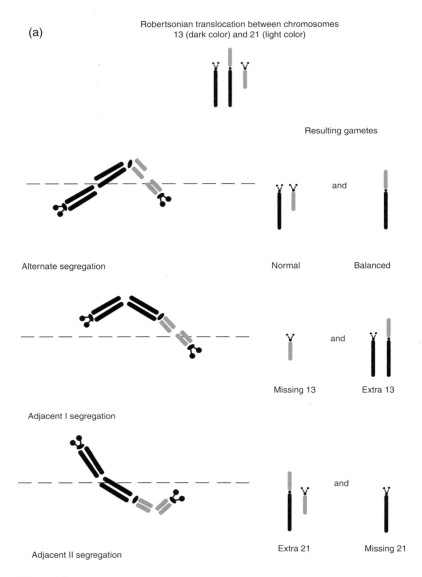

(a)

Robertsonian translocation between chromosomes
13 (dark color) and 21 (light color)

Resulting gametes

and

Alternate segregation

Normal   Balanced

and

Adjacent I segregation

Missing 13   Extra 13

and

Adjacent II segregation

Extra 21   Missing 21

Figure 4.3

(a) The behavior of chromosomes involved in a Robertsonian translocation t(13;21) at segregation during meiosis, and the resulting gametes. (Figure adapted from Connor, M. and Ferguson-Smith, M (eds) (1997) Essential Medical Genetics, 5th Edn. Blackwell Science, Oxford).

either a sperm enters an empty (nullisomic) egg and doubles its chromosome complement or a nullisomic egg is fertilized by two sperm (dispermy). Cytogenetically the karyotype usually looks the same as a normal female 46,XX karyotype or, rarely, the same as a normal male 46,XY karyotype. Maternal uniparental disomy for the full diploid complement results in a benign tumor of the ovary in which embryonic tissues may be present. It occurs in an un-ovulated oocyte, probably when the second meiotic division fails, and chromosomally it shows a normal female 46,XX karyotype.

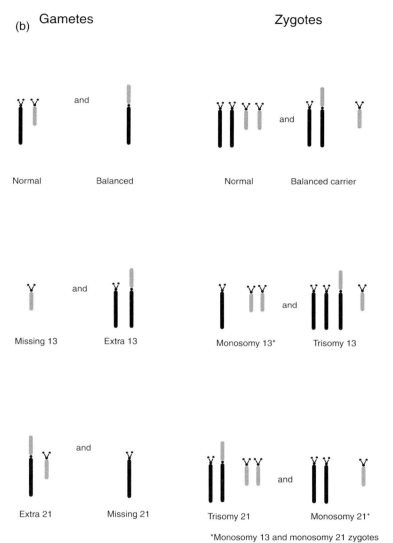

Figure 4.3

(b) When fertilized by normal gametes, the resulting zygotes are also shown. (Figure adapted from Connor, M. and Ferguson-Smith, M (eds) (1997) Essential Medical Genetics, *5th Edn. Blackwell Science, Oxford).*

Uniparental disomy for a whole chromosome occurs when both chromosomes from a homologous pair are inherited from the same parent. It is thought to most commonly result from the loss of a chromosome from an originally trisomic conceptus, thereby restoring it to a disomic state, often referred to as trisomic rescue. If the chromosome lost was the one which was originally duplicated, then the fetus will continue to develop normally. However, if the nonduplicated chromosome is lost then it will result in uniparental disomy for that chromosome. Not all chromosomes are imprinted and so for most there are no obvious adverse phenotypic consequences, unless they happen to be carrying a recessive disease gene. Some, such as chromosome 16, are associated with intrauterine growth retardation, but it is not yet clear whether this is a consequence of the uniparental disomy or a trisomy in the placenta resulting in placental dysfunction. Uniparental disomy of chromosome 15, however, results in either Prader–Willi syndrome or Angelman syndrome, depending on the parental origin of the uniparental disomy. Cytogenetically it is not possible to recognize uniparental disomy of whole chromosomes, although the cytogeneticist may be alerted to the possibility of its presence in prenatal diagnosis when trisomy is present in placental tissue but the fetal cells are disomic.

The diagnosis of uniparental disomy relies on molecular DNA techniques. PCR can be used to compare polymorphisms along the parental chromosome homologs with those observed on the child's in order to determine whether the child has inherited one of each of the parent's chromosomes or only the maternal, or paternal chromosomes.

3.   Nomenclature for human chromosomes

As already mentioned, the human karyotype consists of 22 autosomes and two sex chromosomes. In the construction of the karyotype, the autosomes are numbered according to their size, the largest chromosomes being designated number 1 and the smallest number 22. (Historically, in Down syndrome, the extra chromosome was always designated as chromosome 21, when in fact it is actually smaller than chromosome 22.)

Chapter 3 deals with chromosome banding, but prior to the development of these techniques, chromosomes were assigned groups based on shape and size (see *Figure 4.4*). In the case of breakage studies, where solid staining is still utilized, these categories serve a purpose for designation of the positions of breaks and gaps and other abnormalities associated with these types of studies, as banding may mask small breaks and gaps. The human karyotype can be arranged into seven groups as described below.

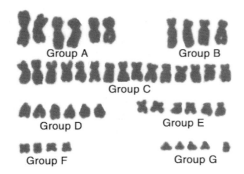

Figure 4.4

Solid-stained metaphase chromosomes arranged in a karyogram, showing the different groups A–F based on shape and size (male karyotype shown).

- Group A – this represents chromosomes 1–3, the largest in the karyotype, with 1 and 3 being metacentric, and 2 being slightly submetacentric. The heterochromatic region near the centromeres of chromosomes 1 and 3 are subject to normal variation in the population.
- Group B – this group represents chromosomes 4 and 5, both large submetacentric chromosomes, which are difficult to distinguish from each other.
- Group C – this group encompasses chromosomes 6–12 and the X chromosome, the largest chromosome in this group. It is almost impossible to distinguish between this group of medium-sized metacentric chromosomes without the use of banding techniques (even with banding techniques they can prove difficult in bone marrow studies). The heterochromatic region near the centromere of chromosome 9 is subject to normal variation in the population (and also inversion).
- Group D – chromosomes 13, 14 and 15 comprise this group, being medium-sized acrocentric chromosomes with satellites and/or short arm formation. These again are difficult to distinguish from one another. The satellites and short arms are subject to normal variation in the population, particularly chromosome 15.
- Group E – chromosome 16, a relatively short metacentric chromosome, is found in this group, and 17 and 18 are relatively short submetacentric chromosomes. One can just about distinguish these chromosomes with practice. The heterochromatic region near the centromere of chromosome 16 is subject to normal variation in the population.
- Group F – chromosomes 19 and 20 are short metacentric chromosomes which cannot really be distinguished without banding techniques.
- Group G – these are the smallest of the chromosomes, including chromosomes 21 and 22, being short acrocentric chromosomes with satellites and/or short arms. The satellites and short arms are subject to normal variation in the population, particularly chromosome 22. The Y chromosome is also placed in this group, as it is usually of similar size to chromosomes 21 and 22, but this can vary widely due to the heterochromatic region on the q arm. If fact in some cases it can almost be as large as a D group chromosome.

The diagrammatic representation of landmarks and bands on the chromosomes was originally described in the Paris conference report based on patterns obtained with Q-, G- or R-banding techniques (Paris Conference, 1971). This serves the cytogeneticist as the standard guide to analyzing human chromosomes, but the one printed in this volume is the author's own interpretation of these ISCN diagrams based on the personal study of many photographs and cells. This was created in an attempt to provide an accurate diagram for teaching purposes with regard to what one actually visualizes down the microscope. Once more it is emphasized that this is *not* a substitute for the ISCN diagrams but can be used to complement that essential volume. The unique ideogram is provided in this volume, in Appendix II, providing the reader with realistic intensities of the G-banding patterns which are observed during analysis.

The length of the chromosome is divided into regions and bands, numbered consecutively from the centromere outward along the p and q arms. The first section nearest the centromere is labeled 1, to the next landmark band, then 2, then 3 and so on. Within each region there are subdivision of bands, 11, 12, 13 and so on. The ends of the p and q chromosome arms are called pter and qter, respectively.

The G-banding pattern shown on the ideograms indicates varying degrees of resolution as shown below:

■ the left-hand version (approximately 400 bands);
■ the middle version (approximately 550 bands), and
■ the right-hand version (approximately 850 bands).

Thus, it can be demonstrated that certain bands appearing as one large dark G-banded area in the lowest banding resolution do in fact split up into two or more areas upon higher resolution. For example, on chromosome 1, band q31 (pronounced 'q three one') can be seen to divide into two smaller dark G-bands. These are then designated subnotations q31.1, q31.2 and q31.3, which include the pale G-banded area (q31.2) revealed by the extension of the chromosome. Logically, one can imagine that the more the chromosomes are extended the more bands are visible. However, in practice the three ideogram representations would suffice as a guide to what one would normally be viewing in practice.

3.1   Nomenclature symbols

Upon first viewing a complex karyotype description, one may be confused by the plethora of evidently strange terms assigned. However, the system for these nomenclature symbols does have some method to its apparent madness. The first part of the karyotype gives the chromosome number, for example 46, a comma is placed after this figure, and the sex chromosomes are recorded, either XX or XY or whatever variations are present. This is followed by another comma (if an abnormal karyotype is being written), and the description is given of the relevant abnormality. The symbols most commonly used are shown in *Table 4.1*. Plus or minus signs are placed prior to the appropriate number to indicate an additional or missing chromosome, for instance a female karyotype with trisomy 8 (one whole additional chromosome 8) would read:

47,XX,+8.

A question mark is used to indicate uncertainty:

47,XX,+?8.

This would indicate that there is an extra chromosome, which is probably chromosome 8. This notation may be required when performing analysis on poor-quality preparations, and very condensed chromosomes, where C group chromosomes can be difficult to identify even when using banding techniques.

A whole missing chromosome (monosomy) would be designated thus, the karyotype showing a male with a missing chromosome 7:

45,XY,−7.

A chimera, or mosaic karyotype, would be listed in numerical order, and not in the order of the most frequently observed cell line, for instance a triple cell line mosaic could read:

45,X [6] /46,XX [4] /47,XXX [10].

Table 4.1 Abbreviations used in written karyotypes

Symbol/abbreviation	Meaning and description
ace	Acentric fragment
add	Additional material of unknown origin
~ (approximate)	Denotes intervals and boundaries of chromosome segment
b	Break
< > (brackets)	(Angled) surrounding a ploidy level figure
[ ] (brackets)	(Square) surrounding a number of analyzed cells
c	Constitutional change
cen	Centromere
chi	Chimera
chr	Chromosome
cht	Chromatid
: (colon)	Denotes break in detailed system
:: (colon double)	Denotes break and reunion in detailed system
, (comma)	Separates chromosome number, sex chromosomes and abnormalities
cp	Composite karyotype
. (decimal point)	Denotes sub-bands
del	Deletion
der	Derivative chromosome
dic	Dicentric
dir	Direct
dmin	Double minute
dup	Duplication
fra	Fragile site
hsr	Homogeneously staining region
i	Isochromosome
idem	Denotes stem line in sub-clones
ider	Isoderivative chromosome
idic	Isodicentric chromosome
inc	Incomplete karyotype
ins	Insertion
inv	Inversion
mar	Marker chromosome
mat	Maternal origin
min	Minute acentric fragment
− (minus sign)	Denotes loss of chromosome
ml	Mainline
mn	Modal number
mos	Mosaic
× (multiplication sign)	Denotes multiple copies of rearranged chromosomes
p	Short arm of chromosome
( ) (parentheses)	Surround structurally altered chromosomes and breakpoints
pat	Paternal origin
pcc	Premature chromosome condensation
pcd	Premature centromere division
Ph	Philadelphia chromosome
+ (plus sign)	Denotes gain of a chromosome

contd.

Table 4.1 contd.

Symbol/abbreviation	Meaning and description
q	Long arm of a chromosome
qdp	Quadruplication
qr	Quadriradial
? (question mark)	Questional identification of a chromosome or chromosome structure
r	Ring chromosome
rcp	Reciprocal
rob	Robertsonian translocation
sce	Sister chromatid exchange
sct	Secondary constriction
sdl	Sideline
; (semicolon)	Separates altered chromosomes and breakpoints in structural rearrangements involving more than one chromosome
sl	Stemline
/ (slant line)	Separates clones
t	Translocation
tr	Triradial
trc	Tricentric chromosome
trp	Triplication
xma	Chiasma(ta)

The numbers shown in square brackets indicate the number of cells observed with each of the differing cell types.

Structural chromosomal rearrangements are specified either by single or three-letter abbreviations of the aberration. The number of the chromosome involved is indicated within parentheses immediately after the symbol describing the rearrangement. If two or more chromosomes are involved, then a semicolon is used to separate them. If a sex chromosome is involved it is always listed first, but all other chromosomes are listed in numerical order. The only exception to this rule is when one chromosome is inserted into another and the chromosome which receives the translocated portion is listed first, regardless of whether it is a sex chromosome or autosome or has a number higher or lower than the donor chromosome. The next set of parentheses indicates the areas where the chromosomes have broken, with the first notation referring to the first chromosome listed, followed by a semicolon and then the next breakpoint referring to the second listed chromosome and so on. Examples are shown below:

$$46,X,t(Y;3)(q12;q21)$$

a translocation between chromosomes Y and 3 where the breaks have occurred in the q arm of Y at band 12 and in the q arm of 3 at band q21.

$$46,XX,t(8;15)(q22;q12)$$

a translocation between chromosomes 8 and 15 with breaks occurring in the q arms of both chromosomes at bands q22 and q12, respectively.

46,XY,dir ins(7;1)(p11;q21q32)

a direct insertion of the portion of chromosome 1 from bands q21 to q32 into a break at band p11 on chromosome 7. If the same portion from chromosome 1 were to be inverted, the karyotype would read thus:

46,XY,inv ins(7;1)(p11;q32q21)

an inverted insertion of the same example above.

When three chromosomes are involved in complex translocations, the sex chromosome is listed first, and the lowest numbered chromosome is listed next. The next chromosome to be listed is the one which receives the portion from the first chromosome, and the last listed chromosome is the one which donates a segment to the first listed chromosome, as shown below:

46,XX,t(9;22;11)(q34;q11;q23)

a complex translocation between chromosomes 9, 22 and 11 where the chromosome 9 segment has moved to chromosome 22, and the chromosome 22 segment has been donated to chromosome 11 and, finally, chromosome 11 has donated its segment to chromosome 9. Note that the chromosomes are *not* in numerical order.

An isochromosome is one where it appears that the chromosome consists of two identical mirror-image arms, either both p arms on either side of the centromere or both q arms. They are designated thus:

46,XY,i(21)(q10)

an isochromosome formed from the q arm of chromosome 21 with the breakpoint being designated as at the centromere q10.

46,XX,i(8)(p10)

an isochromosome formed from the p arms of chromosome 8.

A terminally deleted chromosome would be indicated thus:

46,XY,del(7)(q36)

a terminal deletion at band q36 to qter.

An interstitial deletion would be indicated by two breakpoints in the parentheses following the chromosome number:

46,XY,del(7)(q22q36)

an interstitial deletion of chromosome 7 where the segment q22–q36 is missing. (Note that there is no semicolon between the two regions indicated by the breakpoints as observed in the translocation karyotype.)

Paracentric and pericentric inversions are both indicated by 'inv' but it is evident that 46,XX,inv(2)(q11q13) is paracentric, as the two breaks occur in the same arm whereas, conversely, 46,XX,inv(2)(p11q13) is pericentric as the breaks occur either side of the centromere, in the p and q arms.

Duplicated portions of chromosomes are indicated thus:

46,XY,dup(11)(q21q23)

a duplication of the segment on chromosome 11 from band q21–q23 duplicated at band q21. One can precede the 'dup' abbreviation with 'dir' to indicate a directly duplicated segment, or 'inv' to indicate an inverted segment.

$$46,XY,inv\ dup(11)(q23q21)$$

the same duplication as above but shown inverted.

Ring chromosomes are designated thus:

$$46,XX,r(3)(p21q27)$$

a ring chromosome 3 with breaks at bands p21 and q27. In smaller chromosomes it is not always possible to indicate breakpoints, and in these instances it is admissible to write the notation as:

$$46,XY,r(21)$$

(see Figure 4.5).

A dicentric chromosome is designated thus:

$$46,XX,dic(17)(q11)$$

Breakage and reunion have occurred at band q11 on sister chromatids, to form a dicentric 17.

Derivative chromosomes are designated by the abbreviation 'der', which signifies a rearrangement involving two or more chromosomes, or a number of aberrations in a single chromosome. The abnormal chromosome shown below is designated 'der', being derived from chromosome 5 as it possesses an intact centromere. This example demonstrates an abnormal chromosome 5 formed from an unbalanced translocation between chromosomes 5 and 12. There are two normal chromosome 12s, one normal chromosome 5 and the derivative chromosome 5:

$$46,XX,der(5)t(5;12)(q33;p11).$$

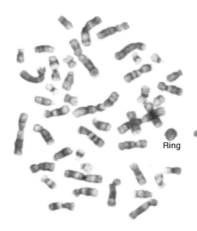

Figure 4.5

Metaphase spread showing a ring chromosome.

More complex changes can also be demonstrated, for example, a derivative chromosome 5 formed from two unbalanced translocations, one involving the short arm, at band p11 with chromosome 7, and one in the long arm at band q33 with chromosome 12, as shown below:

46,XX,der(5)t(5;7)(p11;q22)t(5;12)(q33;p11).

Another example would be a derivative chromosome 7 formed from a deletion of the p arm at band p11 to p13 and an unbalanced translocation at band q22 on chromosome 7, involving chromosome 11:

46,XY,der(7)del(7)(p11p13)t(7;11)(q22;q13).

Dicentric derivative chromosomes can also be demonstrated, as in the example below, showing a break in band q15 in chromosome 6 and in band q22 in chromosome 9, with an additional aberration on the p arm of chromosome 9, where a segment of chromosome 1 has been translocated:

45,XY,derdic(6;9)(q15;q22)t(1;9)(q21;p13).

3.2   Nomenclature symbols for acquired abnormalities and their respective clones

The standard nomenclature terminology for constitutional chromosome abnormalities sufficed to describe those observed in malignant cells for many years, but advances in technology and new data available resulted in some confusion as to the correct designations given in publications. Hence, a revised set of guidelines in nomenclature aimed specifically at acquired abnormalities was published (ISCN, 1991). This has now been updated again (ISCN, 1995), to include the *in situ* hybridization nomenclature and whole chromosome painting. It is not the intention of this chapter to reproduce all the complex nomenclature symbols given, but just to guide the reader through the abbreviations he or she is most likely to encounter in the clinical field.

The modal number is the commonest chromosome number in an abnormal clone. They may not correspond exactly to a haploid, diploid or triploid number and can be expressed for example as near haploid $(n\pm)$ ≤34, hypohaploid $(n-)$ <23, hyperhaploid $(n+)$ 24–34, near-diploid $(2n\pm)$ 35–57, hypodiploid $(2n-)$ 35–45, hyperdiploid $(2n+)$ 47–57, near-triploid $(3n\pm)$ 58–80, and so on. Pseudodiploid or pseudotriploid describes a number equal to any multiple of a haploid number but structural acquired abnormalities are present.

The number of cells that constitutes a clone is given in square brackets [ ] following the karyotype as described in *Section 4.3.1*. Clonal evolution can be shown in the same way as a mosaic in constitutional changes, except that the normal diploid clone, when observed, is always listed last. The mainline (ml) is the most frequently observed cell line in a population of studied cells, which is purely quantitative. As can be seen later (in *Chapter 7*) the amount observed of a particular cell line can be subject to selection bias and the possibility of 'pockets' of certain lines of cells, just as can be observed in histological and morphological analysis of cells.

The stem line (sl) is the most basic clone, the one with the least number of abnormalities, and all other lines are called sidelines (sdl).

46,XY,t(15;17)(q22;q12) [15] / 47,XY,t(15;17)(q22;q12),+8 [5]

This example shows the clone with just the t(15;17) as the stemline (sl) *and* the mainline (ml) as it is observed in the majority of cells. If the sideline (sdl) which shows the t(15;17) and trisomy 8 had been observed more frequently than the line with just the t(15;17), then the sideline would also have been the mainline (ml).

There is sometimes great heterogeneity in solid tumors, but different cells do show some of the same abnormalities and in difficult cases such as these, a composite karyotype (cp) is usually created. An example is shown below:

49–52,XX,del(5)(q13q31),–7,+8,+9,del(11)(q23q25),
+13,add(17)(q23),+20,+21,+22 [cp20].

The composite karyotype gives the range of chromosome numbers of all cells showing the clonal changes and also the range of abnormalities observed. Each of the changes was observed in at least two cells, but there may be no cell with all the changes.

The order of listing chromosome abnormalities is the same as the laws governing the order in constitutional changes. For each chromosome, a numerical change would be listed prior to a structural change, and multiple changes of homologous chromosomes are shown in alphabetical order according to each abbreviation, as shown below:

47,XX,+3,del(3)(q21q26),t(3;5)(q21;q31).

Unidentifiable chromosomes, such as ring chromosomes (r), markers (mar) and double minute chromosomes (dmin) are listed following the karyotype in the above order. Derivative chromosomes whose centromere has not been identified have to be placed following all identified abnormalities, but prior to the r, mar and/or dmin:

47,XX,t(6;11)(q27;q23),+mar

a translocation of chromosomes 6 and 11 with an unidentifiable marker chromosome.

48,XX,t(8;21)(q22;q22),+2mar

a translocation involving chromosomes 8 and 21 with two unidentifiable marker chromosomes.

When several markers are present, they are designated mar1, mar2 and so on, and if multiple copies are present of certain ones, they are designated mar1×3. It should be obvious that, as the marker is unidentified, mar1 does not indicate that the marker is derived from chromosome 1. The karyotype below shows six marker chromosomes, one of which has three identical copies and one has two identical copies, with one occurring as a single copy:

52,XY,t(7;14)(q36;q11),+mar1×2,+mar2×3,+mar3.

Double minutes are special acentric marker chromosomes that should be noted when present in more than one metaphase, abbreviated to dmin. However, they should not be included in the total count, but should be given as a range or mean number per cell.

$$50,XX.......+mar1,+mar2×3,dmin$$

this karyotype shows one double minute per cell.

$$50,XX.......+mar1,+mar2×3,12dmin$$

this karyotype shows a mean of 12 double minutes per cell.

$$50,XX.......+mar1,+mar2×3,6-15dmin$$

this karyotype shows a range of between 6 and 15 double minutes per cell (see *Figure 4.6*).

Acentric fragments (ace) which are not double minutes should not be recorded within the karyotype, for although they may look similar, even if observed in a few cells, they lack a centromere and are usually the result of breakage events and therefore are not clonal.

When a constitutional change occurs within the abnormal acquired karyotype, a letter 'c' following the change indicates a constitutional abnormality. For example, a karyotype with a known trisomy 21 as a constitutional change is observed alongside a karyotype with monosomy 7 and trisomy 8:

$$47,XY,-7,+8,+21c.$$

Additional material of unknown origin is designated thus:

$$46,XY,add(7)(q22).$$

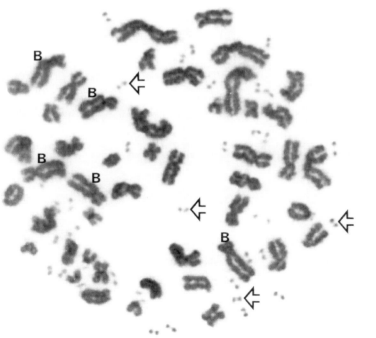

Figure 4.6

Metaphase spread showing double minutes (some double minutes are indicated by the arrows and, as this shows a case with trisomy 4, the five B-group chromosomes are indicated by 'B').

There is additional material on the long arm of chromosome 7 at band q22 of unknown origin.

Depending on where the extra unidentifiable material is located, and also its size, the resulting chromosome may be lengthened or shortened. Obviously additional material at terminally placed bands will result in an increase in chromosome size. However, additional material replacing a chromosome segment (such as that shown in the example above) could lead to an increase or decrease in size, depending on the size on the unknown portion of material. If additional material is present, but it is clear that it has been inserted into a chromosome, the symbol 'ins' must be used rather than 'add' as shown below:

46,XX,ins(3;?)(q21;?).

The symbol 'hsr' is used to describe the presence of a homogeneously staining region (i.e. a region which appears to have no banding pattern) in a chromosome, its arm or a particular band:

46,XY,hsr(19)(p13) or 46,XY,hsr(19)(p11p13)

showing either a homogeneously staining region at band p13 on chromosome 19 or an hsr replacing the region p11–p13, respectively.

The symbol 'idic' is used for isodicentric chromosomes, and 'ider' for isochromosomes formed from a derivative chromosome.

46,XY,idic(12)(p11)

an isodicentric chromosome 12 made up of the long arms of chromosome 12 and the short-arm material from the centromere to band p11.

47,XX,t(15;17)(q22;q21),+ider(17)t(15;17)(q22;q21)

an extra isoderivative chromosome formed from the translocation product of t(15;17).

Multiple copies of rearranged chromosomes are depicted by a multiplication sign. Below, a karyotype is shown with two copies of a deleted chromosome 5, with no normal chromosome 5 being present and the line below shows one normal copy of chromosome 5 but also two copies of a deleted chromosome 5

46,XY,del(5)(q22q31)×2
47,XY,del(5)(q22q31)×2.

48,XY,+del(5)(q22q31)×2

the description above shows two normal copies of chromosome 5 but also two copies of a deleted chromosome 5.

Any uncertainty in band, breakpoint or chromosome designation should be shown by a question mark. For example it is known that the deletion has occurred in the long arm of chromosome 2, in the region q2, but the band within that region is not known:

46,XY,del(2)(q2?).

If it is thought it *may* be in band q21, the designation is written thus:

$$46,XY,del(2)(q2?1).$$

If the question mark precedes the entire abnormality there is a possible deletion in chromosome 2, but everything including the deletion is uncertain as shown below:

$$46,XY,?del(2)(q21).$$

When one cannot be certain about a type of abnormality, for instance a chromosome may appear deleted but also appear like an isochromosome for the p arm, the karyotype can be written as:

$$46,XY,del(7)(q22q36) \text{ or } i(7)(p10).$$

In similar p and q arms such as the chromosome 19, it can be difficult to decide in which of the arms the abnormality has arisen. For instance, if there is extra material, but it cannot be decided whether it is present on the p or q arm, the karyotype can be written as:

$$46,XY,add(19)(p13 \text{ or } q13).$$

When studying malignant chromosomes, the quality of a metaphase spread can be poor, and as such it is difficult to interpret abnormalities, particularly when there are a number of changes in each metaphase, and as such it is not always possible to give a complete picture of the abnormalities present. Here the karyotype is described as incomplete, and the abbreviation 'inc' following the karyotype would denote this. However, this notation should be restricted to the most exceptional of situations, either due to very poor quality or paucity of abnormal metaphase cells.

The new designations used for FISH and whole chromosome paints used in karyotype analysis are discussed in detail in *Chapter 8*.

References

ISCN (1991) *Guidelines for Cancer Cytogenetics, Supplement to An International System for Human Cytogenetic Nomenclature* (ed. F. Mitelman). Karger, Basel.

ISCN (1995) *An International System for Human Cytogenetic Nomenclature* (ed. F. Mitelman). Karger, Basel.

Paris Conference (1971): *Standardization in Human Cytogenetics. Birth Defects* (1972). Original Article Series, Vol. 8, no. 7. The National Foundation, New York. [Also in (1972). *Cytogenetics* 11: 313–362.]

Chapter 5

Approaching analysis

Barbara Czepulkowski

1. Introduction

Understanding what can occur in chromosomal abnormalities that one might come across during analysis is always helpful to the trained cytogeneticist. It is a labor-intensive process which entails the study and recognition of all the banding patterns on each chromosome, and it takes a few years of training to become proficient in dealing with normality let alone abnormality. There are a number of normal variations which have already been discussed (i.e. the heterochromatic variants in the acrocentric chromosomes and those of chromosomes Y and 9) and, as with other disciplines where slides are prepared from human tissue, artifacts of culture must invariably be taken into account.

Analysis may be subjective at times, and there is no laboratory that has not had arguments amongst themselves about the position of breakpoints occurring in abnormalities. When apparent abnormalities occur, usually discrepancies in size and shape, these may be caused by cultural artifacts of a particular slide and a judgment has to be made as to whether the 'change' is real or imagined. Employing the 'eyes' and experience of a trained cytogeneticist is helpful in these situations, although even the best-trained workers can be unsure about small discrepancies. Sometimes you can almost talk yourself into the fact that something is wrong, and at these times it is best to either leave the analysis for a little while, and return to it later, or spread more slides, if material permits this, and study the metaphase cells once more. If the clinician is well versed in the pitfalls of cytogenetics it may be possible to discuss any suspicions with him or her, and perhaps take a further sample if required. Hopefully these types of cases will not occur too often. During the spreading and banding procedures, it should be noted that chromosomes on the outside of the cell can appear to be much larger than those in the center of the metaphase spread. However, the banding pattern should be the same, so this rather than actual size should be taken into account first and foremost.

When studying constitutional changes the approach to analysis differs to that of studying acquired changes. The approach to analysis for acquired abnormalities is discussed in *Section 3*. More often than not, if a change is constitutional, one would usually expect to see it in every cell, so suspicious discrepancies such as the ones noted above should not arise too often. However, there is always the possibility of mosaic cell lines, and as such it is important to keep an open mind during analysis and not make assumptions prior to performing the work. Obviously, the referral form may be helpful, but it should not be used as an excuse to cut corners with analysis if one does not find instantly what is expected. The molecular approaches used to complement nonmolecular cytogenetic techniques can be invaluable in the case of microdeletion syndromes (definitely a case of 'is it?' 'isn't it?' at times!). If the clinical symptoms point to a particular microdeletion syndrome (see *Chapter 6*), it is always best to perform molecular analysis if this is available.

1.1   Methods of analysis

The training provided in individual centers will influence the way every cytogeneticist analyzes chromosomes, but as time progresses each person will settle into their own techniques. Most people are familiar with the picture of regimented chromosomes standing in line, paired up, stuck on a karyotype card as shown in *Figure 5.1*. In practice, this laborious cutting out and sticking down of chromosomes has been superseded by computerized software (not available in every laboratory) and it is only in larger laboratories with good staffing levels that karyotyping in this sense of the word is carried out on each patient. It is usually only for publications and presentations that this practice is still employed in the routine laboratory.

Nearly every cytogeneticist begins life by learning to pair up and recognize chromosomes using photographs, then cutting out the chromosomes, but eventually they should be able to analyze directly down the microscope. When cutting out and pairing chromosomes, three photographs should be printed. One can be used for the cutting out, one for reference (just as you use the picture on a jigsaw box!), and the third photograph can be used for cutting out the crossed-over chromosomes. Use an ISCN ideogram to check on landmark bands and patterns of

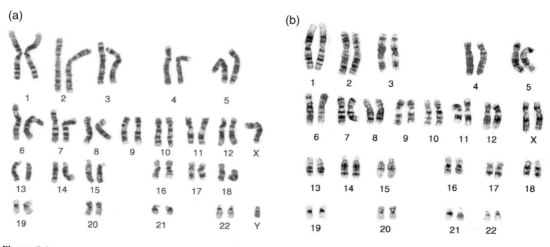

Figure 5.1

Karyogram of (a) human male chromosomes and (b) human female chromosomes, showing a level of banding (approximately 400 bands) which would be beneficial for training the novice cytogeneticist.

chromosomes, or any ideograms, which represent the human karyotype, such as those printed in Appendix II. Eventually, the trainee will become proficient at recognizing the chromosomes instantly, even before cutting them out and pairing them using the photograph, and this is a sure sign that it is time to begin learning the art of analyzing directly down the microscope. When beginning the learning process, it is more sensible to select metaphase cells with a level of resolution such as that shown in *Figure 5.1*, as this is slightly easier than analyzing fully extended chromosomes. This enables the recognition of the landmark bands on the chromosomes. The patterns created by G-banding, the sub-bands and more tangled chromosomes, such as shown in *Figure 5.2*, are the next step. Even locating mitotic cells can be daunting for the novice, and thus a preparation of lymphocytes with a high mitotic index is advisable. It is important to become familiar with the workings of the microscope you are using, to become used to the stage movement, focusing, use of the condenser and, if fitted, the phase contrast. Assuming the novice has been versed in the workings of their microscope, the first steps of chromosome analysis using the microscope can be taken. A low-power objective (×10 or ×16) is the usual way of scanning a slide, and metaphases can be located where chromosomes do not appear too tangled. Using the high-power (×100) oil objective, ensure that the G-banding is of sufficient quality to enable

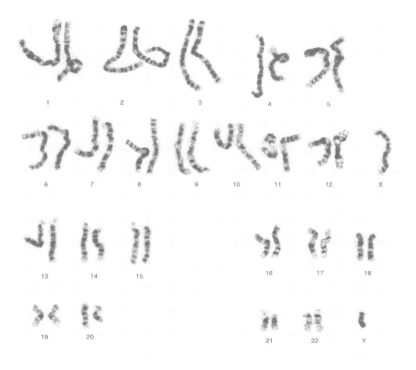

Figure 5.2

Karyogram of human male chromosomes, showing each chromosome paired.

recognition of the chromosomes. Remember that the longer the chromosomes, the more bands are observed and the more tangled they become, hence analysis then appears more difficult to the training cytogeneticist.

Figure 5.3 shows a metaphase spread and some of the different techniques one could use to draw out and pair off the chromosomes. Obviously these are just suggestions, but it must be emphasized, particularly in working with acquired abnormalities, that the first few cells being normal should not lull one into a false sense of security. The author's personal preference has always been to draw a few cells out, and/or have a checklist of the chromosome numbers, ticking each off as you locate them, or writing them down on your drawing or diagram. Mark down or note any queries that may arise so they can be checked out later if required. When any cells are discovered to have chromosomes that do not perhaps conform to the exact shape and/or banding pattern one was expecting, it is usual to return to the cell to re-examine it following further analysis. Because of this,

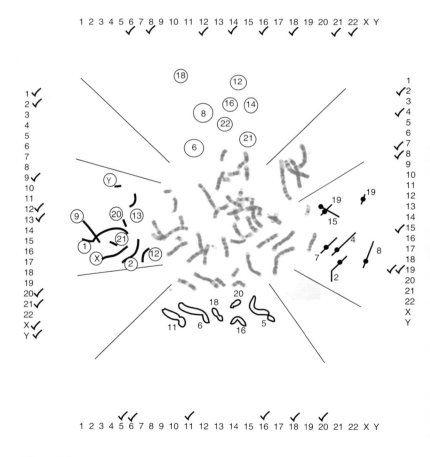

Figure 5.3

Metaphase spread showing various methods of drawing out the metaphase spread when analyzing directly down the microscope.

it is important to make a note of the vernier reading (see *Figure 5.4*) of each cell, and the slide which has been used, by numbering the slide in question and writing this, in addition to the reading, in the analysis book. The more information that is documented, the easier it is to return to the cell to make a fresh assessment. There are two vernier scales on the stage of the microscope and both readings should be taken to establish the coordinates of the cell in question.

An alternative to the vernier reading is the England finder (see *Figure 5.4*), which is a glass slide marked over the top surface with letters and numbers in a grid pattern so that a reference position can be deduced by direct reading. Reference numbers run horizontally 1–75 and letters vertically A–Z (omitting I). A point of interest on a slide is noted and brought into the center of the field. The slide is then carefully removed, and

VERNIER READING SHOWING 25.2

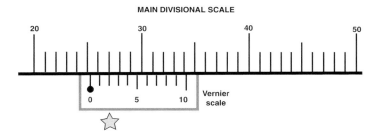

Note that the gray star indicates where the lines from the Main divisional scale and the smaller Vernier scale match up, giving the figure beyond the decimal point.

ENGLAND FINDER SCALE
READING P31/4

Note that the gray star indicates the point of interest, making the England Finder reading P31/4.

Figure 5.4

Vernier reading and England finder reading.

the England finder is placed upon the stage (there are left and right markings on the England finder, also an 'up' arrow) under the microscope under low power. A reading is noted down, and this enables any microscope user to find the metaphase in question, even on different microscopes, because each microscope has a different vernier scale.

The number of cells analyzed varies in each laboratory and is also dependent upon the workload involved. For constitutional analysis, *Section 2* describes the suggested minimal number for the various types of referral categories. Common sense would indicate further study of more metaphase spreads if a problem cell is discovered, and this is not unknown in studying normal as well as abnormal cells. Random changes occur, which are not clinically significant, and an odd translocation or other change in only one cell would not constitute an abnormality. The cytogeneticist would carry out further checks on other cells before making a judgment on one abnormal cell. It should be noted that, when peripheral blood is analyzed, if the patient has had a viral infection many strange and random changes in the chromosomes can be observed, perhaps in more than one cell.

Quality of banding is crucial for very subtle abnormalities such as microdeletions, and NEQAS, following slide assessment from laboratories, has produced guidelines for the G-banding quality required depending upon the referral category (*Guidelines for Clinical Cytogenetics UK National External Quality Assessment Schemes (NEQAS*, 1994). The scoring figures are based on G-banding quality, and also the quality of spreading on a slide. As has been shown above, random chromosome loss and tangling and overlapping chromosomes where one cannot distinguish particular areas of a chromosome, can adversely affect the accuracy of analysis. If these problems are apparent, points are deducted and, as such, the preparation may not reach the required standard. It is an unwritten 'law' that with the microdeletion you are seeking, the chromosome region you wish to study is always hidden under another chromosome, obscured by it, or even missing! *Tables 5.1* and *5.2* are adapted from the NEQAS scoring guide.

2.   Guidelines for the analysis of chromosomes from constitutional studies

2.1   Chromosome changes occurring from conception to term

Cytogenetic abnormality is the most common cause for the majority of abnormal conceptions being lost between fertilization and 12 weeks. From the chorionic villi, cardiac blood, cord blood, amniotic fluid, skin and placenta, five cells analyzed and five counted is the minimal requirement. If a mosaic is observed, a further 20 cells should be counted.

Chromosome abnormalities which can be observed during this period include:

- autosomal trisomies;
- 45,X Turner syndrome;
- other sex chromosome abnormalities;
- triploidy;
- large unbalanced rearrangements;
- small unbalanced rearrangements;

Table 5.1 G banding quality scoring criteria

Points/description	Description of banding and landmark bands required
0/No banding	Unequivocal chromosome pairing is not possible. This sometimes occurs if the staining is too dark, or the cells are simply resistant to attempts at banding
2/Poor	Approximately 150 bands observed per haploid set. Although fine detail cannot be established, unequivocal pairing can be achieved, e.g. 8 and 9 or 4 and 5 should be distinguishable from each other
4/Moderate	Approximately 400 bands observed per haploid set. Examples of this quality include two distinct bands on the 8p and 9p, and three distinct bands at 5q14, 5q21 and 5q23
6/Good	Approximately 550 bands observed per haploid set. Examples of this quality include three bands on 11p, and four distinct bands on 18q. Resolution should allow 7q35 to be visible, and also 22q13.2
8/Excellent	Approximately 850 bands observed per haploid set. Examples of this quality include resolution enabling 6q24, 6q25.2 and 6q26 to be seen as three distinct bands, 11p14.1 and 11p14.3 can be seen, 15q12 should be distinct and 20p should have at least two dark bands

Table 5.2 Recommended minimum quality for referral categories

Reason for referral	G-banding
Prenatal diagnosis, e.g. age or biochemical prescreens	2
Aneuploidies and known large structural rearrangements	2
Expected small structural rearrangements, including their prenatal diagnosis	3
Possible small unknown structural anomaly, e.g. recurrent abortion, dysmorphic features, delayed development, mental retardation	5
Microdeletion syndromes	7

- microdeletions;
- unbalanced Robertsonian translocations.

As described in *Chapter 4*, on types of abnormalities, it is clear that gametes from patients with balanced translocations can be unbalanced and as such would give rise to unviable fetuses. Multiple spontaneous abortions would alert a clinician to the possibility of such a translocation.

2.2   Chromosome changes occurring in neonates and young children with congenital malformations and/or developmental delay

Individuals with multiple congenital abnormalities would be suspected of having a chromosomal defect, and cytogeneticists should familiarize themselves with the clinical symptoms of the various recognized syndromes (see *Chapter 6*). Again, five cells analyzed and five counted is the minimal requirement for these categories and if any numerical or structural abnormality is suspected, 20–30 more cells should be counted and/or examined. Chromosome changes which occur in the above categories could include:

- trisomy 13 Patau syndrome;
- trisomy 18 Edward syndrome;
- trisomy 21 Down syndrome;
- unbalanced Robertsonian translocations;
- small unbalanced inherited structural rearrangements;
- *de novo* deletions;
- 45,X Turner syndrome;
- microdeletions;
- ring chromosomes;
- Fragile X.

Sometimes there is no apparent chromosomal change on first inspection, so areas of chromosomes which may be suspected of having a microdeletion because of the accompanying symptoms should be studied in greater detail. However, the most efficient way of discovering if a microdeletion is present is by the new molecular techniques which are described in *Chapter 8*, and any suspicion which is aroused should be tested molecularly if possible. In addition, the chromosome change may be extremely subtle and as such may remain undetected until further investigations are made using, for example, the new chromosome painting techniques.

2.3   Chromosome changes occurring during puberty

Although sex chromosome abnormalities can be suspected and subsequently disclosed prior to puberty, it is at this stage, when failure or partial failure of normal secondary sexual development occurs that most sex chromosome abnormalities are suspected and investigated. The clinical features of these sex chromosome abnormalities are detailed in *Chapter 6*. Five cells analyzed and at least 25 cells counted is the minimal number required in this category. A larger number of cells is required because mosaics are commonplace. The chromosome changes occurring here include:

- 45,X Turner syndrome;
- abnormalities of X chromosomes;
- XXY Klinefelter syndrome;
- XY females;
- XX males;
- XYY.

2.4   Other chromosomal observations

Marker chromosomes

This could be a whole subject and chapter alone, as the origins of markers can be wide-ranging and diverse. Many are derived from rearrangements in the pericentric regions of acrocentric chromosomes, especially chromosomes 15 and 22. They can also be derived from small rings or centromeric fragments of other chromosomes. Look out for the possibility that acrocentric chromosomes naturally attract each other via the phenomenon of satellite association, so this could point to a marker from an acrocentric. However, don't expect this sort of help all the time!

The marker can be bi-satellited, have satellites on one end or even no satellites. The smaller bi-satellited markers usually have no euchromatic material and as such cause no phenotypic effect to the individual, so these are the types of markers which can be familial. However, those with significant amounts of euchromatin are usually *de novo* markers, and can sometimes produce a number of phenotypic effects. The most commonly observed reciprocal translocation in man is the t(11;22)(q23;q11) translocation, and the minute der(22) can be observed in karyotypes as an extra marker.

To establish the origins of markers, C-banding, Q-banding, NOR staining and DA-DAPI staining may be applied. Employing centromere-specific probes or chromosome paints for anything suspected to be a euchromatin marker is also helpful.

Heterochromatic variants in the normal population

With practice and experience, the detection of normal variants will not be a cause for dread for the cytogeneticist, although it is still possible to encounter large variations, which may cause the cytogeneticist to think there is an abnormality. What may appear to be abnormal could in fact be just a normal variation, and these can be quite common. In these cases it is essential to differentiate from normal variants and structural rearrangements. C-banding should be attempted on any suspicious variants by novice cytogeneticists, until using G-banding alone the cytogeneticist has seen enough variants to feel confident that no further testing is required (which may take many years). Any unexpected or unusual findings should be checked using the parental karyotypes, if possible. As described in *Chapter 4*, the normal human karyotype may show variations in the short arms and satellites of acrocentrics. Variations at the heterochromatic region near the region of chromosomes 1, 3, 9 and 16 are also observed in the normal human karyotype and a problematical area is chromosome 9, which can also be inverted. In fact some individuals show two inverted chromosome 9s.

The Y chromosome is normally the size of a G-group chromosome, but can have such variation that it may almost be the size of a D group, so the trainee cytogeneticist should beware. C-banding, Q-banding and Y heterochromatic probes can all be used to confirm the heterochromatic nature of such variants. Extremely small Y chromosomes are seen which have little or no heterochromatic material. The easiest and best way to investigate the heterochromatic nature of this sort of variant is to look at the paternal Y. If this does not match, then a deletion of Yq has occurred.

Reporting heterochromatic variants is not advisable, as the referring clinician could well interpret this as an 'abnormality'. This laboratory does not report these normal variants, but if they are reported, it must be made completely clear in the report that these are of no clinical significance and using the phrase 'normal variation' is advised.

2.5   Interpreting 'normal' and abnormal karyotypes

'Normal' is not the best description on a report, as this may indicate that there is no defect at all. This may not be the case, as subtle changes could be beyond the resolution of the microscope, as is the case in some microdeletion syndromes. The cell type can also be important, as the abnormal cells may not be present in the tissue being studied for cytogenetic analysis. With the case of trisomy 20, and isochromosome for 12p, only skin and other tissues demonstrate this change but not the lymphocytes, which is, as mentioned previously, the cell type one normally studies for cytogenetics. Problems are also encountered when normal cell lines outgrow the abnormal lines in fibroblast cultures. These are all problems which the cytogeneticist must be aware of when analyzing any case. When reporting any such case, it is usually best to say 'no abnormalities detected', which may include a multitude of sins. These problems apply to any tissue studied, including chorionic villi and amniotic fluid, and hence sometimes unexpected phenotypic abnormality may be apparent when the infant is born following a 'no abnormalities detected' report. Clinicians and patients alike should be made aware of the pitfalls and limitations of any such test.

As is outlined in detail in acquired changes in *Section 3*, one must never jump to conclusions having detected an abnormality, no matter how tempting this may be. Although in constitutional analysis additional abnormalities are not often observed, they can occur and dramatically alter the interpretation and recurrence risks. The study of structural rearrangements can be enhanced by observing both balanced and unbalanced forms using family studies.

If translocations and other structural changes are detected, they should be followed up. In these instances it is usual to study parental karyotypes in order to establish if the translocation is inherited. Family studies may be useful. The carriers of such a balanced translocation would, as mentioned previously, be at risk of spontaneous abortions, despite the ability to produce normal offspring either with or without the translocation.

Trisomies such as those of chromosomes 21, 18 and 13 are usually *de novo* and, unless a Robertsonian translocation is observed, for instance t(13;21), it is unlikely that family studies will prove informative.

Deletions, such as the terminal deletion of chromosome 5p (Cri-du-chat syndrome), usually arise *de novo* and parental karyotypes are usually normal.

It is essential to fully explain any abnormality on a report, using the ISCN conventions, but also describing what has been seen. We have found that a diagram (ideogram) of the abnormality and/or a photograph is usually appreciated by the referring clinician.

Some 'balanced' rearrangements occurring *de novo* will have some form of genetic damage, and a small proportion of phenotypically abnormal individuals will have this type of apparently 'balanced' rearrangement.

2.6   Problems of interpretation arising in cell culture

Due to the nature of cell growth, in various types of culture techniques, different problems can arise when studying chromosomes in these situations. When cells are grown *in situ* (see *Chapter 2*), growth is initiated from each cell that settles on the substrate, and in turn forms a colony of cells. Visible colonies are usually the result of the growth of an initial cell. On finding a numerical or structural abnormality in a colony, examination of other cells in the colony should be sufficient to demonstrate that the abnormal cell is an artifact if only one such cell is discovered. It is possible that more cells in the same colony could show an identical abnormality, but if one checks and examines the surrounding colonies and these prove to be normal, then once more it can be assumed the abnormal cells are probably an artifact of culture. There is an approximately 3% rate of mitotic errors during *in situ* culture and, as such, artifacts should be expected. Where trisomies, sex chromosome abnormalities, triploidy and markers in a single colony are detected, as many colonies should be examined as are available and, although this type of change may represent true fetal mosaics, most laboratories do not report this type of finding to clinicians. However, further investigation on suspicious cases, such as ultrasound studies and fetal blood sampling, should be employed if there is any doubt whatsoever.

The nature of cell culture does not preclude the possibility that a spontaneously arising abnormal cell colony becomes fragmented and gives rise to a number of other colonies. Luckily, when multiple colony abnormalities occur with autosomes, instead of the usually seen translocations and/or inversions, the abnormalities tend to be odd and unusual, with ring chromosomes, isochromosomes and small segments of the ends of euchromatic chromosomes occurring. Common sense would indicate that any uncertainty or suspicion of a particular abnormality should be followed up with further studies as mentioned above. The commonly observed trisomies of 21, 13 and 18 should always be investigated further. Trisomy 20 is a headache, as fetal blood sampling may not demonstrate the presence of this abnormality even though it may be a true fetal abnormality. Ultrasound may prove informative, but true mosaics with trisomy 20 may appear phenotypically normal.

The very nature of suspension harvests makes a chance encounter with an abnormal cell even more difficult to interpret. Any abnormality in a suspension harvest would warrant at least a further 30 or so cells being examined if possible. When known abnormalities such as trisomies, triploidy or markers are observed in more than one cell in a suspension harvest, then not only should further cells be examined and checked, but parallel cultures which should have been initiated should also be investigated. Suspicion of true fetal abnormality would obviously arise if the same chromosomal abnormality were found in parallel cultures, in the same way that multiple colonies with the same abnormality in *in situ* cultures would be a cause of concern.

Direct and long-term cultures are usually set up in most laboratories from chorionic villi tissue and both cultures should be studied if possible prior to a final report. Although it is tempting to report the presence of autosomal trisomies of 21, 13 and 18 on direct cultures alone, occasionally a normal fetus can be a possible outcome from the alleged 'abnormal' direct culture. Any sex chromosome abnormalities found in both the direct and cultured cell is usually a true finding. It is still prudent to use fetal blood sampling or amniocentesis as an additional check even in these cases. As for *in situ* chorionic villus cultures, any mosaic findings of trisomies for chromosomes 21, 13 or 18 should be followed up by amniocentesis or fetal blood sampling.

The problem of maternal cell contamination in cell culture

The hazard of maternal cell contamination comes up for all prenatal culture work as it is observed in 0.5% of long-term cultures from amniotic fluid and chorionic villi. The unexpected discovery of female cells when one has already studied a predominantly male culture is usually the result of maternal cell contamination. However, one must always try to exclude the possibility of a human error laboratory mix-up, which is not unknown. Substandard samples (i.e. blood-stained ones), should be noted in the day-book prior to set-up. Samples that are blood-stained upon receipt often give rise to poor growth in culture (see *Chapter 2*) and, with very few colonies available for examination and assessment, may in fact turn out to be maternal cells. The Kliehauer test on all blood-stained samples is helpful, as this distinguishes between fetal and maternal cells. If most of the cells prove to be maternal, then suspicions may arise if a female culture results from such samples. It is usually prudent to quote the 0.5% risk of maternal cell contamination of such samples on the written report.

In the case of chorionic villus cultures, the cytogeneticist may already be wary of some long-term cultures, as samples that have a large amount of contaminating maternal decidua upon receipt may prove unreliable if strict cleaning procedures are not adhered to during the setting up process. Again a figure of 0.5% of cultures could show maternal cell contamination, but this is dependent upon the skill of the cytogeneticist in removing any contaminating maternal decidua during the setting up of these types of culture. This is where the direct cultures prove useful, as they are usually thought to carry the correct sexing for an individual. A dilemma may arise if there is an abnormal female cell line alongside a normal cell line, as this could indicate either a true fetal abnormality, a mosaic form of that abnormality, or a true fetal abnormality with maternal cell contamination. This type of problem requires further follow-up studies as described above.

Confined placental mosaicism

This is another problem which faces the cytogeneticist when studying chorionic villus cultures. Mosaic forms of trisomies other than 21, 13 and 18 are usually confined to the placenta, and mosaic sex chromosome abnormalities are also unreliable with regard to the true karyotype of the fetus. However, marker chromosomes in either mosaic or nonmosaic form may be genuine in about half of the cases that are studied.

Trisomy 16 may occur as a true fetal abnormality, and gives rise to congenital abnormalities. As such it should be investigated further if noted in chorionic villus cultures. It is unfortunate, however, that nearly every other trisomy has been reported in the literature around the globe somewhere, but unless trisomies for chromosomes 8, 9, 15, 16 and 22 are observed, other trisomies are usually considered as cultural artifacts.

Other changes

Unexpected structural rearrangements should be investigated using parental karyotypes as one would do when such an abnormality is found in any prenatal culturing procedure.

For a more extensive description of interpretation of constitutional changes, the reader is referred to Wolstenholme and Burn (1992, 2000).

3.   Guidelines for the analysis of chromosomes from malignant tissues

The suspicion that chromosome abnormalities were present in neoplastic tissues sparked the explosion in studies of this type of tissue. However, abnormalities are not detected in all malignant cell cultures. In fact in some cases genetic changes have occurred, but only at a molecular level, which are impossible to detect at the resolution of the eye following normal G-banding procedures. An example of this is the *BCR/ABL* gene rearrangement where the visible t(9;22) translocation is not detected in a suspected case of chronic myeloid leukemia by G-banding. It is also a possibility that not all neoplastic cells have resulted from a genetic rearrangement. It must always be stressed to the clinician that a 'normal' karyotype does not mean that the disease is not present in the tissue being examined. Neoplastic cells coexist with normal cells, hence most abnormal cases also show normal metaphases, and some apparently normal cases will show a few abnormal cells if a sufficient number of cells is examined. Although the presence or absence of karyotypically normal cells may be of some prognostic significance, the proportion of normal to abnormal cells is subject to varying culture conditions, and probable sampling error.

This problem is even further complicated for the cytogeneticist by the fact that increased proliferation of any normal cells present in a sample may be favored by certain culture regimes, or even delays in transport of the sample. The results one achieves when analyzing malignant material are dependent on the factors shown in the following sections.

3.1   Prompt arrival of the sample

Following removal of a bone marrow sample from the patient and subsequent transport, any normal cells present will generally proliferate preferentially. If the sample can be transported to the laboratory as quickly as possible and placed in the refrigerator prior to setting up, or stored in the refrigerator prior to transportation, this will provide near-optimum conditions for detecting any abnormal clone which may be present.

If, however, the center requesting the cytogenetic test is far removed from the laboratory, the sample is often posted, and obviously no refrigerator will be encountered during the sample's journey through the postal system! Normal results from these types of samples should be regarded with caution, and delays in receipt must be noted and the clinician informed. In these instances, it is advisable to analyze a larger number of cells than the standard used by the recipient laboratory.

3.2   Metaphase quality and availability

The chromosomes of neoplastic cells are often of poor morphology and as a result banding may also be substandard. Thus, if an inexperienced cytogeneticist inadvertently selects the best metaphase spreads on the slide, only the cells with a normal karyotype may be detected. It is important to encourage the trainee cytogeneticist to analyze anything and everything! A distinctly mixed population of spreads on the slides should be a warning that a detectable abnormality, if present, is more likely to be found in the poorest quality cells. No matter how tempting it is to ignore the poor quality cells, it is necessary to analyze them.

3.3   The number of metaphase spreads to be analyzed

One would imagine that the more spreads analyzed, the greater the chance of detecting an abnormal clone. There are tables available which show the number of cells that must be analyzed to exclude mosaicism at various confidence limits (Hook, 1971).

Depending upon the workload imposed on a clinical laboratory and material available, 30 cells fully analyzed is probably the best criterion to adopt. However, a heavy workload often reduces this to between 15 and 20. An analysis of less than 10 cells is probably inadequate unless of course an abnormality has already been detected in those 10 cells.

If low-level mosaicism is expected, such as in the case of a patient where there is a suspicion of relapse of a condition where a previous abnormality has been detected, extended analysis and examination is advisable, and between 50 and 100 cells should be examined if possible. Although cytogenetic analysis can be useful for detecting minimal residual disease, other molecular techniques are probably more useful. A vast number of cells can be screened in, for example, PCR, which is 100 000 times more sensitive than nonmolecular cytogenetic techniques. For patients entered in treatment trials, there are certain requirements laid down by the organizing body, and so the appropriate number of cells should be analyzed in these particular cases.

When a specific abnormality has been identified in even 10 cells, it would be acceptable to examine further cells for this change without full analysis of the other chromosomes. This approach would also be appropriate where certain clones have already been identified in varying degrees of progression such as 46,XX,t(8;21)/45,X,−X,t(8;21)/ 45,X,-X,t(8;21),del(9). In such a case, the proportion of each clone could be established quite quickly. Normal cells coexisting with an abnormal clone can also be detected using this technique. This is particularly true when examining a culture for minimal residual disease, and it is obvious if the first few cells do not show the previously detected abnormality.

If one chromosome causes suspicion, such as a chromosome 7 in a case of MDS?/ AML?, one would obviously examine a large number of cells, if possible, in order to rule out the possibility that the suspicious chromosome may have been a result of a cultural artifact. The more cells available for analysis in these cases the better.

3.4   Cytogenetic analysis experience

The detection rate of abnormalities varies widely in different centers and sometimes may be explained by the experience of the cytogeneticists in-house. It can take even an experienced cytogeneticist some time to get used to the variable quality observed when analyzing bone marrow samples. This is due to the poor quality of metaphase spreads which is often encountered, and the need to analyze the worst-looking cells (and consequently the most challenging and time-consuming cells) on the slide, as these may well conceal the abnormality. It is also a great advantage to have observed the abnormalities previously, and to have recognized the derivative counterparts in a translocation. Being aware of the appearance of the abnormal chromosomes within a particular translocation, or even other structural changes such as deletions and inversions, is of great benefit when untangling the bizarre and complex karyotypes that may occur when studying chromosomes in disease. *Chapter 7* provides the reader with some examples of some of the more common changes observed in analysis of malignant tissue. The abnormality may exist in a plethora of other changes, such as in the case of myelodysplastic syndromes (MDS) transformed or transforming to acute myeloblastic leukemia (AML). This phenomenon is also observed in well-advanced cases of acute lymphoblastic leukemia (ALL), where the unbalanced form of a structural abnormality can be concealed amongst a number of other changes. Another scenario such as this occurs when studying bone marrow samples from patients with lymphoma, because by the time the disease has infiltrated the bone marrow (following origins in the lymph node) there may be many secondary changes masking the important and sometimes diagnostic primary change.

This may sound like an obvious statement, but counting the cells is also most important, as it is surprising how many cytogeneticists, in their inexperience, have missed something truly simple such as trisomy 8. Even drawing out the cells is sometimes not foolproof, as it has been known for the metaphase to be drawn with the correct number of chromosomes, but the extra chromosome to be left devoid of a label. Some workers draw and tick off each chromosome as they go along and sometimes this can cause problems, as when the analyzer, say, finds two chromosome 8s, that is it! (The extra chromosome 8 sometimes remains not drawn out, or unlabeled.) Due to the nature of the study of chromosomes in malignancy, checking may seem futile in some cases, as one is usually searching for an abnormal clone amongst a number of normal cells. Hence, the onus is on the person who originally picks up the sample to make a careful study of any metaphase spreads available. The advantage of having an experienced eye is that a subtle chromosome change is more likely to be detected by an 'old hand' rather than a 'new recruit'. However, even the best analyzer may fall foul of a very subtle change, particularly if the quality of the sample is compromised in some way. Due to the very nature of the request for rapid detection of certain changes in the clinical environment, for example the t(15;17) in the case of acute promyelocytic leukemia (APML), banding is sometimes performed when the slides are 'hot off the press'. Quality is obviously diminished in these cases, and one would not risk a solitary slide in these instances. However, if there is material remaining following processing of the sample, then a positive or negative answer for the presence of certain diagnostic translocations as soon as possible may have huge prognostic implications for the patient, and greatly aid their management.

The advent of FISH, and all forms of chromosome painting and molecular analyses, has proved extremely useful in sorting out complex changes or even picking out subtle changes such as the t(12;21) in ALL, which had not been detected until a few years ago (Romana *et al.*, 1993).

4.   Clonal abnormality

Clonal abnormality has been defined by conventions laid down by the Fourth International Workshop on Chromosomes in Leukemia 1982 (1984) and an abnormal clone is said to exist if:

■ two or more cells have the same structural abnormality;
■ two or more cells have acquired the same chromosome (trisomy);
■ three or more cells have lost the same chromosome (monosomy).

When chromosome abnormalities occur in single cells, they should generally be ignored, as random changes do occasionally arise in all types of cells. This is particularly true if the patient has received some form of chemotherapy and the bone marrow is being analyzed following that therapy, as random changes do occur following treatment.

However, if a single cell shows an abnormality which is known to be associated with a particular disease type [for instance t(9;22) in a case of suspected or known chronic myeloid leukemia (CML)], this should not be completely ignored. A repeat sample, if possible, would however be advisable. If they remain as single-cell findings, they should be reported with an explanation, but require confirmation before being considered as a true abnormal clone. Discussion with the referring clinician is the usual procedure in such cases. Exceptions to this could be made in the case of a highly specific abnormality such as the t(15;17) in a case of known or suspected APML. If it is not feasible to take a further sample, molecular studies on interphase cells (material permitting) should be undertaken.

An apparently balanced chromosome rearrangement not commonly associated with the disease type being examined should alert the cytogeneticist to the possibility of a constitutional anomaly, particularly if this is observed in every cell. Studying the karyotype of a PHA-stimulated peripheral blood sample should rule out or confirm the presence of a constitutional change, which would then probably not be clinically significant for the disease being studied.

The interpretation of chromosomal loss must be approached with great caution. The smaller chromosomes, in particular, are often lost due to spreading artifacts. It is important to note whether a particular slide has been subject to overspreading, with a large number of incomplete metaphase spreads, prior to drawing any conclusions about a particular lost chromosome. If there is any doubt about the significance of an apparent monosomy, it is wise to scan a large number of cells before giving any interpretation. Even if convention states that three or more cells should have lost the same chromosome to constitute monosomy for that chromosome, apparent monosomy is often observed with the small chromosomes, such as chromosomes 21, 22, 19 and 20. This may occur in culture simply due to artifacts of spreading. Examine the rest of the slide to see if there is loss from a number of cells, and if in doubt either re-spread the slides or count more cells.

When studying follow-up samples from patients who have undergone some form of chemotherapy and/or transplant, obviously it is hoped that the abnormal clone detected at diagnosis has disappeared. In the case of CML patients with the t(9;22) translocation, conventional therapy did not have any positive impact upon the abnormal cells. However, clinical trials with interferon therapy are showing some success in certain patients, by irradicating the cells with t(9;22). At present, 100% removal of the abnormal clone is still not as frequent as one would hope, most patients showing variable responses to the therapy.

The detection of abnormality in neoplastic cells is not as easy as it sounds and can be extremely difficult, particularly in cases of acute leukemia where one would expect a change but none is detected. However, with an educated application of both culture technique and also analysis, satisfaction can be had in performing these tests. A discussion of the more common abnormalities observed in malignant disease can be found in *Chapter 7.*

References

Fourth International Workshop on Chromosomes in Leukemia 1982 (1984) *Cancer Genet. Cytogenet.* **33**: 254.

Guidelines for Clinical Cytogenetics UK NEQAS (1994) Potten, Barber and Murray, Bristol.

Hook, E.B. (1971) *Am. J. Hum. Genet.* **29**: 94.

Romana, S.P., Le Coniat, M. and Berger, R. (1993) *Genes Chromosomes Cancer* **9**: 186.

Wolstenholme, J. and Burn, J. (1992) The application of cytogenetic investigation in clinical practice. In: *Human Cytogenetics: a Practical Approach*, Vol. I. (eds D.E. Rooney and B. Czepulkowski). IRL Press, Oxford, pp. 119–156.

Wolstenholme, J. and Burn, J. (2000) The application of cytogenetic investigation in clinical practice. In: *Human Cytogenetics 3e: Constitutional Analysis* (Vol. 1), Practical Approach Series no. 240. Oxford University Press, Oxford.

Chapter 6

Constitutional chromosome abnormalities

Barbara Czepulkowski and Karen Saunders

I.   Introduction

Approximately 20% of all conceptions in humans have a chromosomal disorder and, although this figure appears high, most of these conceptuses do not implant or are spontaneously aborted. In early spontaneous abortions, the frequency of chromosomal aberration is 60%, whereas in later spontaneous abortions and still births, this figure drops to 5%. The types of abnormalities observed include trisomy and 45,X triploid or tetraploid changes. Each type of autosomal trisomy has been observed, with the exception of chromosome 1, with trisomy 16 being particularly frequent. The most common chromosomal disorders and their frequencies observed in newborns are shown below:

Pericentric inversion	1 in 100
Balanced translocation	1 in 500
Trisomy 21 (Down syndrome)	1 in 700
47,XXY (Klinefelter syndrome)	1 in 1000 males
47,XXX	1 in 1000 females
47,XYY	1 in 1000 males
Unbalanced translocation	1 in 2000
Trisomy 18 (Edward syndrome)	1 in 3000
Trisomy 13 (Patau syndrome)	1 in 5000
45,X (Turner syndrome)	1 in 5000 females

Autosomal chromosome abnormalities cause more severe phenotypic effects than sex chromosome changes, and deletions cause more problems than duplications. With autosomal changes, mental handicap, multiple congenital abnormalities, growth retardation and dysmorphic features are common recognized symptoms.

Translocations which are balanced should cause no effects to health and lifespan, but carriers of such a translocation would be subject to suffering spontaneous abortions, as during meiosis abnormal gametes are formed from the abnormal translocation products (see *Chapter 4*). Normally the zygote with an unbalanced translocation will spontaneously abort, although some do go to term and if live born will show multiple dysmorphic features and mental handicap. The risk of producing live born unbalanced offspring depends upon the type of translocation and which parent is the carrier. For instance if a centric fusion has occurred between the same chromosomes [e.g. t(21;21)] in either parent, the risk of producing unbalanced offspring is 100%. However, with a centric fusion between two different chromosomes [e.g. t(13;14)] in either parent, the risk is just 1% (see *Figure 6.1*). Exchanges of whole arms of metacentric and submetacentric chromosomes will rarely produce viable offspring, and such changes are only observed via studies on spontaneous abortive material. The new approaches to prenatal diagnosis have allowed us to observe the karyotype of the conception at an earlier stage, using chorionic villus sampling (see *Chapter 2*). As a result of this new technique, the chance of unbalanced offspring for a reciprocal translocation carrier is about 23%, whereas half of this percentage would have aborted by the time amniocentesis was performed. Hence, using chorionic villus material, one is more likely to observe such unbalanced offspring when karyotyping is performed.

Family studies should be carried out in order to establish other outwardly healthy carriers who could be at a similar reproductive risk. When a *de novo* balanced translocation has been detected where both parents have normal chromosomes, it is not usually associated with clinical abnormality. In less that 10% of *de novo* reciprocal balanced translocations, genes may be damaged at the breakpoints, which can give rise to an abnormal phenotype, and these instances are always problematical when a *de novo* translocation is an otherwise incidental finding at amniocentesis.

Paracentric inversions do not produce clinical abnormality in the carrier, nor do they warrant an indication for prenatal diagnosis, because if a crossover occurs within the inversion loop it would usually be incompatible with viability.

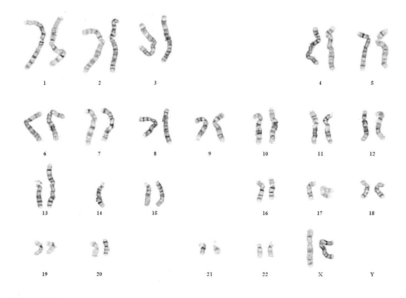

Figure 6.1

Karyogram of a Robertsonian translocation involving chromosomes 13 and 14. (Photograph supplied by Angela Douglas of the Liverpool Women's Hospital.)

Table 6.1 Common terms used in dysmorphology

Term	Meaning
Brachycephaly	Short antero-posterior skull length
Brushfield spots	Speckled iris ring (20% of normal babies)
Canthi	Inner margins of eye
Clinodactyly	Incurved fifth fingers
Coloboma	Missing segment of iris or eyelid
Craniosynostosis	Fused sutures (prenatally)
Dolichocephaly	Long antero-posterior skull length
Epicanthic folds	Skin folds over inner canthi
Holoprosencephaly	Abnormal development of forebrain
Hypertelorism	Interpupillary distance above expected
Hypotelorism	Interpupillary distance below expected
Low set ears	Upper border of ear attachment below intercanthal line with head upright
Macrocephaly	Large head, associated with hydrocephalus and hydraencephaly
Microcephaly	Small head, usually mentally retarded
Micrognathia	Small chin
Nystagmus	Involuntary rapid movement of the eyeball
Prognathia	Jutting chin
Ptosis	Drooping upper eyelid
Simian crease	Single palmar crease
Telecanthus	Inner canthal distance above expected but inter pupillary distance not increased

Pericentric inversion carriers (where the centromere is involved in the inversion) who are not clinically or phenotypically affected are still at risk of producing unbalanced offspring, especially if the inversion involves a large portion of the chromosome. The risk is 8% for a female carrier and 4% for a male carrier. The exception to this is the normal variant of inversion of chromosome 9, within the heterochromatin near the centromere, which is present in about 1% of the population.

The following sections describe common chromosomal disorders with their clinical features. *Table 6.1* explains the meaning of some common terms used in dysmorphology.

2.   Trisomy 21 – Down syndrome

The incidence of live births is shown above and increases with advancing maternal age, and at amniocentesis (16 weeks) the incidence is 1 in 300 for a 35-year-old mother, whereas this rises to 1 in 22 at 45 years of age.

Ninety-five per cent of cases are those arising from nondisjunction at the first but sometimes second meiotic division. It is the mother who contributes the extra chromosomes in most cases (85%), and the father in 15%. As was noted in the above section, some patients can be mosaic, that is have normal and abnormal cell lines which may arise from mitotic nondisjunction in a trisomy 21 zygote or occasionally in a normal zygote. About 4% of cases arise from a *de novo* translocation, or the extra copy of chromosome 21 has arisen through a balanced translocation involving this chromosome.

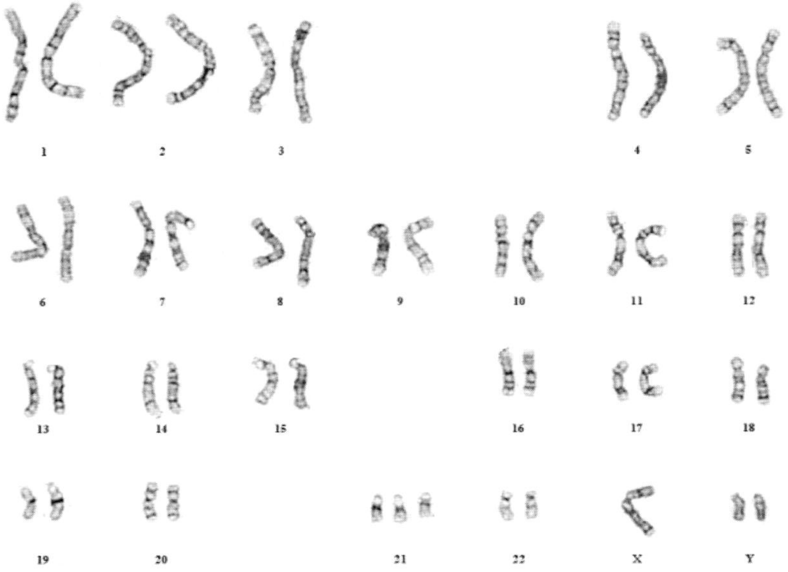

Figure 6.2

Karyogram demonstrating trisomy 21 (Down syndrome) and an extra Y chromosome. (Photograph supplied by Angela Douglas of the Liverpool Women's Hospital.)

2.1   Clinical findings

The facial appearance usually alerts the clinician to the possibility of Down syndrome. The palpebral fissures are upslanting, the eyes have Brushfield spots (a speckling of the iris), the nose is small, and the facial profile is flat. In the newborn, there may be marked hypotonia with redundant folds of skin around the neck. The skull is brachycephalic, and the ears are misshapen and low set. There may be a Simian crease (a single palmar crease), and clinodactyly may also be present. A serious problem is mental handicap as the IQ is usually found to be less than 50. If this is not so, it is usually because the individual is a mosaic. Congenital heart defects may be present, and death in infancy is common. Clinical complications may include cataracts, epilepsy, hypothyroidism, acute leukemia and atlantoaxial instability. Trisomy 21 accounts for about 25% of all moderate and severely mentally handicapped children of school age. Puberty is often delayed, and pre-senile dementia often sets in after 40 years of age.

In young patients, where a trisomy 21 or mosaic for trisomy 21 child has been produced, the risk of recurrence is 1.5%. However, if the mother is over 35 years, the age-specific risk should also be taken into account. Offspring have been known to be born to female trisomy 21 patients, and less than half are affected (see *Figure 6.2*).

3.   Trisomy 18 – Edward syndrome

This abnormality usually arises due to maternal nondisjunction, mainly at the first meiotic division. Paternal nondisjunction may also occur but is much rarer. The severity of the symptoms is reduced in an individual who is mosaic for this trisomy. The older a mother is, the more the chances of producing a trisomy 18 increase. The incidence at conception is quite high, but 95% of the affected zygotes abort spontaneously. There is a higher percentage of female trisomy 18 newborns, possibly due to more male trisomy 18 conceptions aborting spontaneously.

3.1 Clinical findings

The birth weight is usually low, with multiple dysmorphic features. Organ malformations (heart and kidney) are common, with 30% of live births dying within a month. There is gross developmental delay with only 10% surviving the first year. There is a small chin and prominent occiput, low-set malformed ears, clenched hands with overlapping index and fifth fingers, a single palmar crease, rocker-bottom feet and short sternum. In males, cryptorchidism is usually present.

If the parents have a regular trisomy 18 child, the risk of recurrence and/or other major chromosomal changes is 1.5% at amniocentesis.

4. Trisomy 13 – Patau syndrome

Trisomy 13 arises mainly from maternal nondisjunction, particularly in the first meiotic division. Again, paternal nondisjunction is known but less frequent. In about 20% of cases, it is revealed that one of the parents is a translocation carrier, and mosaicism occurs in about 5% of patients.

4.1 Clinical findings

There is usually congenital heart disease, and 50% of newborns die within a month. Gross developmental delay is usual and only 10% survive the first year. Dysmorphic features are present, including hypotelorism (reflecting underlying holoprosencephaly), microphthalmia, cleft lip and palate, scalp defects, ear abnormalities, redundant nape skin, clenched fists, a single palmar crease, polydactyly and prominent heels. In males, cryptorchidism is usually present.

The recurrence risk is below 1% unless one parent is a carrier of a balanced translocation.

5. 45,X – Turner syndrome

The missing X chromosome can arise from nondisjunction in either parent. In 80% of Turner syndrome patients, only the maternal X is present, therefore the error would have arisen during spermatogenesis, or following fertilization. Types of X abnormality giving rise to Turner syndrome are shown below:

45,X	50%
i(Xq)	17%
Mosaic	24%
r(X)	7%
del(Xp)	2%

Short-arm deletion of the X chromosome generally gives rise to the Turner phenotype, while deletions of the long arm of chromosome X give rise to streak ovaries, without the accompanying dysmorphic features. Mosaicism is occasionally observed with a second cell line, which contains a Y chromosome, and in these instances there is a 20% chance of the streak ovaries developing a gonadoblastoma. The vast majority of conceptions (99%) spontaneously abort.

5.1   Clinical findings

Redundant neck skin is often present in the newborn, and peripheral lymphedema, although it is common for the abnormality to be identified later following investigation for short stature or primary amenorrhea. The short stature is noted from childhood and there is no adolescent growth, the mean height for an adult being 145 cm. A broad chest is often present, giving the appearance of widely spaced nipples. There may be webbing of the neck and a low hairline. The fourth metacarpals may be short, and hypoplasia of the nails and multiple-pigmented naevi are common. Ovaries tend to develop normally until about the 15th week of gestation, but then a degeneration occurs, giving rise to streak ovaries at birth. Secondary sexual characteristics fail to develop during puberty.

Other symptoms can include congenital heart disease, and an increased risk of systemic hypertension. Renal abnormalities may be present, as well as Crohn's disease and gastrointestinal bleeding.

The intelligence and lifespan are usually normal, and sex hormone replacement therapy can result in the development of secondary sexual characteristics. Growth hormone treatment in early childhood has been shown to counteract the short stature, but only minimally. Risk of recurrence is not above normal population levels.

6.   47,XXY – Klinefelter syndrome

This abnormality arises by nondisjunction at the first or second maternal meiotic division and rarely as a mitotic error following fertilization. Paternally, the abnormality arises when an XY sperm is produced at the first meiotic division. About 15% of the affected individuals are mosaic for 46,XY/47,XXY.

6.1   Clinical findings

Diagnosis is normally made following investigations for infertility, and is the single most common cause of hypogonadism and infertility in males. The testes are small (less than 2 cm long) in the adult, and testosterone levels are low. Secondary sexual characteristics fail to develop fully due to the low testosterone levels, and gynecomastia is present. The limbs are long, with a mean adult height in the 75th centile. Other problems include scoliosis, emphysema, osteoporosis, diabetes mellitus, and the frequency for breast carcinoma is similar to that for normal females. Severe mental handicap is rare. The risk of recurrence is not above normal population levels (see *Figure 6.3*).

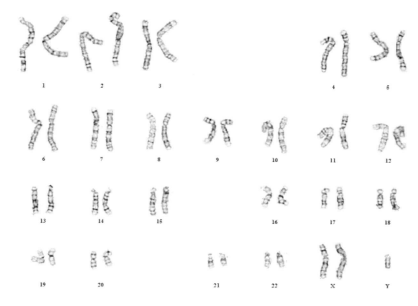

Figure 6.3

Karyogram demonstrating Klinefelter syndrome (XXY). (Photograph supplied by Angela Douglas of the Liverpool Women's Hospital.)

7.   46,XX males

46,XX males arise from an accidental recombination of the short arms of the X and Y chromosomes in paternal meiosis. As a result of this, transfer of chromosome Y sequences occurs, including testis-determining factors from the Y to the X chromosome. One in 20 000 males have an apparently normal female karyotype, but on extended chromosomes banding studies reveal transfer of Yp11.2 to Xp and, in most other cases, DNA analysis or *in situ* hybridization identify the presence of Y-specific sequences.

7.1   Clinical findings

These 46,XX males are sterile, with endocrine features of Klinefelter syndrome (see *Section 6*) and small testes. Intelligence is usually normal, and stature within the normal female range. Diagnosis is usually made during investigations for infertility or when prenatal diagnosis predicts a female infant but the newborn is an apparently normal male. Risk of recurrence is not above normal population levels.

8.   47,XYY

This abnormality arises from the production at second paternal meiotic division of a YY sperm, or nondisjunction of the Y chromosome following fertilization.

8.1 Clinical findings

47,XYY is normally asymptomatic although intelligence may be reduced compared with other siblings and behavioral problems may arise. Patients do tend to be tall, but are proportional. Fertility is unimpaired and, although one would expect the possibility of the following offspring, 2 XXY/2 XY/ 1 XX/1 XYY, only XX and XY offspring have been observed.

The risk of recurrence is probably not increased following the production of an affected individual.

9. 47,XXX

This abnormality may arise from nondisjunction at either first or second maternal meiotic divisions, nondisjunction at the second paternal meiotic division or a post-zygotic error.

9.1 Clinical findings

Clinically, these individuals appear normal although there is a chance of mild mental handicap. Three-quarters of affected individuals are fertile; one-half of their offspring would be expected to have the same chromosome abnormality but in fact their offspring are usually normal. Risk of recurrence is not above normal population levels.

10. Triploidy

Triplody involves an extra haploid set of chromosomes, giving rise to trisomy for every chromosome in the complement. It is usually paternally derived, 66% being the result of double fertilization, 24% due to fertilization with a diploid sperm, and 10% due to the fertilization of a diploid egg. From this it can be seen that most (66%) would be 69,XYY. Hydatidiform change occurs only when there is a double paternal contribution (see *Chapter 4, Section 2.3*). Triplody occurs in about 2% of all conceptions, but survival to term is extremely rare, and most are spontaneously aborted.

10.1 Clinical findings

Multiple congenital abnormalities and a low birth weight is apparent in the newborn. There is also a very small trunk in proportion to the head size, syndactyly, a large placenta and hydatidiform changes.

The recurrence risk is unknown but probably no higher than that for the normal population.

11. Microdeletion syndromes and chromosomal duplications

There is an increasing number of clinical syndromes, caused by microdeletions and occasionally microduplications. These may arise via a new mutation or at meiosis where the parent carries a balanced structural rearrangement. If a visible chromosomal imbalance is detected, it is inevitable that an abnormal phenotype results, usually manifesting itself as multiple dysmorphic features and/or mental retardation. Usually, the experienced physician guesses that a chromosome disorder may be the cause of the problems, even though it would not be possible to tell which chromosome is involved. Embryonic development and brain function are two processes which depend on the greatest number of genes, therefore these functions are most vulnerable to changes in gene dosage. It is interesting to note that the phenotypes have striking similarities regardless of which chromosome is involved. *Table 6.2* below gives a list of some of the microdeletion syndromes.

The presence of these abnormalities can be confirmed by gene dosage studies, which can also aid in gene localization. The smallest visible deletion involves 4000 kb and from this it can be seen that large numbers of genes can be lost or even gained without producing detectable cytogenetic changes. In fact it is only through molecular studies that certain single gene disorders have been recognized. It is not within the scope of this book to list the molecularly detectable changes, suffice to say that around 6000 have been described which affect around 2% of the population (Connor and Ferguson-Smith 1997; see *Figure 6.4a* and *Figure 6.4b* in the color section between pages 18 and 19).

11.1 Prader–Willi and Angelman syndromes

These syndromes (the clinical features of which are described in *Table 6.1*) occur when there is a loss, or a loss of function, of the genes located on the long arm of chromosome 15 at q11–13. This most commonly occurs due to a microdeletion of this area. However, a loss of function of these genes may also occur due to uniparental disomy (UPD; see *Chapter 4, Section 2.3*) of this region. Prader–Willi syndrome results from deletion of the paternally derived chromosome region (70% of cases), or loss of the paternally inherited critical region due to maternal UPD of chromosome 15 (about 28% of cases). Angelman syndrome results from deletion of the maternally derived chromosome region (70% of cases) or loss of the maternally inherited critical region due to paternal UPD of chromosome 15 (about 4% of cases). The remaining cases of Prader–Willi, and about one-third of the remaining Angelman syndrome cases are due to absence or mutation of the imprinting center in the critical region, which leads to nonfunctioning of the Prader–Willi/Angelman genes. The rest of the Angelman cases are thought to be due to a mutation of a structural Angelman syndrome gene (ASHG/ACMG report, 1996; see *Figure 6.5* in the color section between pages 18 and 19).

12. Marker chromosomes

A small additional marker chromosome of unknown origin is observed in about one in 2500 pregnancies. These can prove a testing problem for the cytogeneticist, although 90% are derived from the short arms, and centromeric regions of acrocentric chromosomes and about half of these again are usually from chromosome 15. Using DAPI staining (see *Chapter 3, Protocol 3.8*), it is possible to establish whether chromosome 15 is involved in the additional marker material. Occasionally one of the parents carries

Table 6.2 Microdeletion syndromes

Chromosome region	Syndrome	Clinical features
1q32-q41	Van der Woude (Syndrome 1)	Paramedian pits on lower lips, and cleft lip or palate.
2p21	Holoprosencephaly 2	Bilateral mid-line cleft lip, hypotelorism.
2q34-q36	Van der Woude (Syndrome 2)	Paramedian pits on lower lips, and cleft lip or palate.
3p25-pter	3p25-pter deletion syndrome	Cerebellar hemangioblastoma, retinal angioma, renal carcinoma.
3q22-q23	Blepharophimosis-Ptosis-Epicanthus inversus	Hypertelorism with marked epicanthus, ptosis.
3q24-q25	Dandy–Walker	Hydrocephalus, malformation of 4th ventricle of brain.
4p16.3 (deletions and translocations)	Wolf–Hirschhorn	'Greek warrior helmet' appearance, microcephaly with frontal bossing, high forehead with large glabella, hypertelorism, broad nose, micrognathia, short upper lip.
4q12-q13	Piebald trait	Hypopigmented areas, mainly ventral surfaces of trunk and face, photophobia, and nystagmus.
5p15 (deletions and translocations, but submicroscopic in minority)	Cri -du-chat	Characteristic 'mewing' cry in infancy, microcephaly, round face, hypertelorism, micrognathia, epicanthic folds, broad depressed nasal bridge.
7p21-p22 (53% deletion some translocations)	Saethre–Chotzen	Broad flat nasal bridge, telecanthus, maxillary hypoplasia, craniosynostosis, mental retardation and syndactyly.
7p13 (deletions and translocations in 50%)	Greig cephalopoly syndactyly (GCPS)	Frontal bossing, scaphocephaly, hypertelorism, broad first digits, pre and postaxial polysyndactyly.
7p11.2-p14	Craniosynostosis (syndrome 1)	Prominent frontal bossing, occular hypertelorism, broad flat nasal bridge and large mandible.
7q11.2	Williams	Short stature, mental handicap, hypercalcaemia, supravalvular aortic stenosis, and characteristic facies, prominent lips and small nose.
7q11.23 [deletions in proximal breakpoint in two cases of inv(7)]	Zellweger	High forehead, expressionless face, micrognathia and shallow supra orbital ridges, hypotonia, hepatomegaly, retardation, death before 6 months.

contd.

Table 6.2 contd.

Chromosome region	Syndrome	Clinical features
8p11.1	Spherocytosis type II (ankyrin defect)	Spherical red cells, hemolytic anemia, splenomegaly.
8q11-q13	Brachio-oto-renal	Pre-auricular pits, cup shaped ears, brachial sinuses and fistulae, deafness, renal hypoplasia.
8q24.11 (deletions q23.3-q24.13, and inversions)	Trichorhinophalangeal (TRP1)	Sparse hair, lax skin.
8q24.11 (deletions q24.11-q24.13)	Langer–Giedion (TRP2)	Moderate mental retardation, bulbous nose with tented alae, large philtrum, thin lips, sparse hair, lax skin, cartilaginous exostoses.
9q21.2-q23		Mental retardation and growth delay
10q11.2	Hirschsprung's disease	Absence of submucosal and myenteric ganglion cells on rectal biopsy, the cause of neonatal constipation and abdominal distension.
11p15.5 (duplication paternal; translocations part of p15 – maternal)	Beckwith–Wiedemann	Macroglossia, anterior abdominal wall defect, high birth weight, ear distal lobe grooves and hemihypertrophy.
11p13 (deletions and balanced translocations)	Wilm's tumor (WT), WAGR – (WT, anaridia, genital anomalies, mental retardation)	Absent irides, ptosis and photophobia, lens opacity and glaucoma, genital abnormalities and mild mental retardation, kidney tumors.
13q12.2	Moebius	Bilateral facial weakness, and bilateral abducens palsy.
13q14 (deletions visible in 3–5% cases, translocations and point mutations)	Retinoblastoma	Malignant tumor of the eye in children usually occurring in the first 2 years, may be unilateral (sporadic, 60–70%) or or bilateral (may be inherited 30–40%). 5% of the bilateral cases have the deletion of 13q14.
15q12 (maternal) (60–70% show deletions at q12)	Angelman (AS)	'Happy puppet' syndrome. Developmental delay, poor speech, jerky movements, paroxysms of laughter, dysmorphic facies.
15q12 (paternal) (70% show deletions at q12)	Prader–Willi (PWS)	Newborn-hypotonia, flat face, tented upper lip, hypoplastic genitalia. Childhood obesity, prominent forehead, small hands and feet, low IQ.
16p13.3	Alpha-thalassemia/mental retardation (ATR-16)	Features depend on which α globin genes absent, ineffective erthyropoiesis, aggregation of impaired α globin chains, anemia.

contd.

Table 6.2 contd.

Chromosome region	Syndrome	Clinical features
16p13 (submicroscopic deletions in 25%)	Rubinstein–Taybi (RSTS)	Broad terminal phalanges, thumbs and halluces, mental handicap, abnormal facies.
17p13.3 (90% show deletion in LIS gene)	Miller–Dieker (MDS)*	Lissencephaly-agyria-smooth brain, microcephaly, prominent occiput, high forehead, with narrow mid face, low set ears.
17p11.2 (duplication of PMP-22 gene in 65–85%, or point mutation)	Charcot–Marie–Tooth 1A	Generalized muscle wasting particularly in lower limbs.
17p11.2	Smith–Magensis	Brachycephaly, mid face hypoplasia, prognathia, hoarse voice, speech delay, mental and growth retardation, hyperactivity.
20p11.23 (deletions in some cases)	Alagille	Peripheral pulmonary artery stenosis, biliary hypoplasia, and variable liver impairment, prominent forehead, deep set eyes, and variable retardation.
22q11.21-q11.23 (deletions in 15–20%)	DiGeorge 2/CATCH 22	Neonatal seizures, thymic aplasia, failure to thrive, aortic arch anomalies, dysmorphic facies with hypertelorism, down slanting palpebral fissures and fish-like mouth.
Xp21 (some deletions visible, submicroscopic in 50–60% of affected boys or point mutations)	Duchenne muscular dystrophy (DMD)	Progressive proximal muscle weakness in early childhood, walking delayed, elevated serum creatine kinase, absence of dystrophin in muscle.

*The Miller–Dieker syndrome MDS abbreviation is *not* the same as the MDS abbreviation used for myelodysplastic syndromes

the additional marker chromosome and this may indicate that phenotypically it should not affect the child, although this is not always the case. *De novo* marker chromosomes are even more of a problem and almost a whole chapter could be dedicated to this problem alone. However, the reader is referred to *Human Cytogenetics: A Practical Approach*, where this subject is dealt with in greater detail (Wolstenholme and Burn, 2000).

13. Ring chromosomes

Ring chromosomes, as described in *Chapter 4*, result from breakage in both arms of a chromosome with subsequent fusion of the two arms at the points of breakage to form a ring. The chromosome segments distal to the breaks are therefore lost, resulting in a partial monosomy for these areas. Clinically this usually causes severe dysmorphism and mental retardation in

(a)

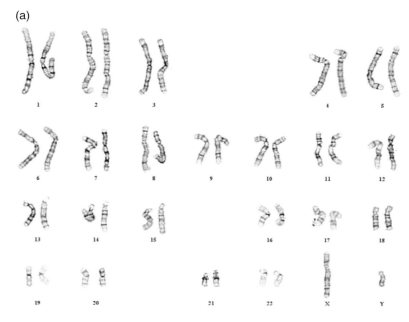

Figure 6.4

(a) Karyogram demonstrating the microdeletion of chromosome 22 at band 22q11.2 (the abnormal chromosome on the right – notice how difficult it is to detect). (Photograph supplied by Angela Douglas of the Liverpool Women's Hospital).

the carrier. Occasionally the chromosome arms fuse practically at the telomeres with no loss of coding DNA and these individuals may be phenotypically normal. However, some of these individuals also show phenotypic abnormality due to a disruption of cell division during fetal growth, caused by the ring becoming tangled or broken. This can give rise to partial or total aneuploidy, in some cells, for the chromosome involved in the ring. This mosaicism may show characteristics which include growth retardation, slight to moderate mental retardation and mild dysmorphic features. The larger chromosomes are the most prone to this effect since the ring is more likely to become tangled and is therefore less stable during cell division.

14. Fragile X syndrome

The Fragile X syndrome is associated with moderate to severe mental retardation and a folate-sensitive fragile site band at Xq27.3. It occurs in 1 in 4000 male and 1 in 10000 female births (deVries *et al*, 1998). The name is derived from the region FRAXA located in the 5′ untranslated region of the FMR1 gene. The gene encodes for the protein product FMRP, which is required for normal brain development. Three states can exist where the untranslated region houses a CCG triplet repeat: normal (6–55 copies), pre-mutation (55–230 copies) and full mutation (230–1000 copies). Males and females with the pre-mutation are normal, males with the full mutation have Fragile X. In females, where there are two X chromosomes present, the situation is more complex. Those with the full mutation will have a variable mental impairment, and the level of mental retardation is usually correlated to the proportion of Fragile X-positive cells observed. Intellectually normal females may express a very low level of Fragile X cells, or not at all.

The clinical features of patients include autism, hyperactivity or just mental retardation. There may be a prominent jaw, forehead and large ears. Following puberty (in boys), macroorchidism is common.

As the *FRAXA* site is folate sensitive, it can be induced by preparation in a medium with a low folate content, or with folate antagonists such as fluorodeoxyuridine or methotrexate, and also adding excess thymidine will have similar effects (see *Chapter 2*). A fragile site would be observed in 4–60% of cells in affected males, whereas carrier females may show a small percentage of cells, but half of them show no fragile site expression. These methods were used traditionally, and one can see that they would be liable to produce false negative results, particularly when studying females. It was also necessary to screen large numbers of cells in order to detect the presence of Fragile X cells, and as a consequence these methods were extremely laborious.

Thankfully, the methods have been superseded by Southern blot analysis, which establishes the size and methylation status of the CCG repeat in *FMR1*.

There is also a second, much rarer, folate-sensitive fragile site on the X chromosome that is thought to be associated with mild mental retardation. This second site, known as *FRAXE*, lies distally to the *FRAXA* site at *Xq28* (Sutherland and Baker, 1992). Cytogenetically it is induced in the same way, and the resulting fragile site is indistinguishable from that of *FRAXA*. It also involves an expansion of a CCG triplet repeat. Normal individuals are thought to have 6–35 copies of this triplet repeat, whereas individuals with mental retardation and a fragile site expressed cytogenetically possess more than 200 copies. This results in methylation of the adjacent CpG island at the 5′ untranslated region of the *FMR2* (Fragile X mental retardation-2) gene and a silencing of its transcription (Chakrabarti *et al.*, 1996). However this fragile site has been much less studied than *FRAXA* and the exact clinical implications surrounding it have yet to be determined since there seems to be considerable clinical variation between families and individuals with the mutation. *FRAXE* is detected by Southern blot analysis, using a protocol similar to that used to detect *FRAXA*.

Acknowledgments

We are indebted to Angela Douglas of the Liverpool Women's Hospital for providing the excellent photographs of constitutional abnormalities for this chapter. For these we are extremely grateful.

References

ASHG/ACMG Report (1996) *Am. J. Hum. Genet.* **58**: 1085–1088.
Chakrabarti, L., Knight, S.J.L., Flannery, A.V. and Davies, K.E. (1996) *Hum. Mol. Genet.* **5**: 275–282.

Connor, M. and Ferguson-Smith, M. (eds) (1997) Single gene disorders. In: *Essential Medical Genetics* (5th edn), pp 130–149. Blackwell Science, Oxford.

deVries, B.B.A., Halley, D.J.J., Oostra, B.A. and Niermeijer, M.F. (1998) *J. Med. Genet.*, **35**: 579.

Sutherland, G.R. and Baker, E. (1992) *Hum. Mol. Genet.* **1**: 111–113.

Wolstenholme, J. and Burn, J. (2000) In: *Human Cytogenetics: a Practical Approach*, 3rd edn. (eds D.E. Rooney and B. Czepulkowski). Oxford University Press, Oxford.

Further Reading

Connor, M. and Ferguson-Smith, M. (eds) (1997) *Essential Medical Genetics* (5th edn), Blackwell Science, Oxford

Chapter 7

Acquired chromosome abnormalities observed in malignancy

Barbara Czepulkowski

1. Introduction

The discovery of chromosome abnormalities in malignancy was difficult to interpret in the clinical situation prior to G-banding. It was noted many years ago that malignant cells often showed alterations in chromosome number and structural changes. However, it was impossible to place any clinical significance on these changes, as discerning which chromosome was involved in any particular abnormality was impossible prior to banding. In the 1960s, a small acrocentric chromosome was nearly always observed associated with chronic myeloid leukemia (CML) and, as it was discovered by Nowell and Hungerford (1960) in Philadelphia, it was given the grand title of the Philadelphia chromosome (see *Figure 7.1*). In the 1970s, following the introduction of G-banding (Seabright, 1972), which allowed the recognition of each separate chromosome, it became clear that this small chromosome was in fact the result of a rearrangement, a translocation, between chromosomes 9 and 22. Despite this revelation, this abnormality is still commonly known as the Philadelphia chromosome.

Following this discovery, the study of chromosomes in malignant tissues became more common, and recurring chromosome abnormalities in

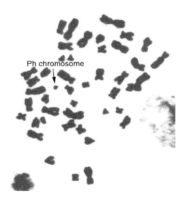

Ph chromosome

Figure 7.1

Solid-stained preparation showing the small Philadelphia chromosome (arrowed) as Nowell and Hungerford (1960) may have first observed it.

particular disease types were observed, which in turn were associated with these diseases. The following chapter will discuss the more common abnormalities observed in disease, with particular reference to the acute leukemias. The changes observed were nonrandom, and molecular studies have also begun to show that structural changes caused genes to be either disrupted or deregulated and expression was altered, thus giving rise to abnormal proteins, and consequently abnormal cell development. If one studies the blood cells, it is clear that different types of leukemia correspond to failure in development at the various maturation stages, that is acute promyelocytic leukemia demonstrates a failure of the white cells to progress beyond the promyelocyte stage. The specific translocation involved (see *Section 3.5*), the t(15;17), has been studied molecularly and it has been demonstrated that the genes *PML* on chromosome 15 and *RARα* on chromosome 17 are disrupted. Treatment with *trans*-retinoic acid, as opposed to other acute myeloid leukemia therapies, has improved the prognosis for these patients dramatically. It is hoped that further discoveries regarding other changes involved in the varying types of leukemia will lead to the same improvement in long-term survival.

Missing (monosomy) and extra (trisomy) chromosomes would also have a similar genetic effect, although the mechanisms would be different to that observed in diseases which demonstrate a translocation. One could consider an overdose, so to speak, of a particular gene in trisomy and a missing gene effect in monosomy. Deletions could be subject to the missing gene from the lost segment not having the desired affect on the gene on the other homolog, giving rise to abnormalities in development. Interstitial deletions could also have the same effect as a translocation, as genes not normally next to one another, are either disrupted by the break or placed in close proximity to an alternative gene may have an unwanted effect on the development of the cell.

Some abnormalities are extremely specific, such as the t(15;17), whereas other changes such as trisomy 8 is considered to be a myeloid change with no real specificity apart from the indication that an abnormal clone is present in the malignant myeloid cells. The reasons for the differences in this type of specificity are not easy to explain, and these questions are still unanswered.

2. Classification and clinical symptoms of myeloid malignancies

2.1 Acute myeloid leukemia

Hematological diseases were originally classified using morphological features, supplemented by immunophenotyping. The French American British (FAB) classification [Fourth International Workshop on Chromosomes in Leukaemia, 1982, (1984)] of acute myeloblastic leukemia is shown in *Table 7.1*.

Cytogenetics has now also been included in categorizing these conditions and *Table 7.2* shows some abnormalities with their FAB types and molecular changes.

Table 7.1 FAB classification of acute myeloblastic leukemia

FAB type	Description	Incidence (%)
M1	Myeloblastic without maturation	15–20
M2	Myeloblastic with maturation	30
M3	Promyelocytic (hypergranular)	5–10
M3V (variant)	Promyelocytic (hypo- or micro-granular)	5–10
M4	Myelomonocytic	15–20
M4Eo	M4 with eosinophilia	15–20
M5a	Monoblastic	15
M5b	Promonocytic or monocytic	15
M6	Erythroblastic	3–4
M7	Megakaryoblastic	2–4
M0	Myeloblastic with minimal differentiation	<1

Table 7.2 FAB categories with specifically associated cytogenetic abnormalities

FAB category	Cytogenetic abnormality	Molecular change
M0, M1 or M2	t(9;22)(q34;q11)	*BCR/ABL* fusion
M2	t(8;21)(q22;q22)	*AML/ETO1* fusion
M2Baso	t(6;9)(p23;q24)	*DEK/CAN* fusion
M3 or M3V	t(15;17)(q22;q21)	*PML/RARα* fusion
M3 like	t(11;17)(q23;q21)	*PLZF/RARα* fusion
M4	t(1;11)(q21;q23)	*MLL/AF1q* fusion
	t(10;11)(p13;q14)	*CALM/AF10* fusion
M4Eo	inv(16)(p13q22) and t(16;16)(p13;q22)	*CBFβ/MYH11* fusion
M4 or M5	t(6;11)(q27;q23)	*MLL/AF6* fusion
	t(8;16)(p11;p13)	*MOZ/CBFβ* fusion
	t(11;19)(q23;p13.1)	*MLL/ELL* fusion
	t(11;19)(q23;p13.3)	*MLL/ENL* fusion
M5	t(1;11)(p32;q23)	*MLL/AF1p* fusion
	t(4;11)(q21;q23)	*MLL/AF4* fusion
	t(10;11)(p12;q23)	*MLL/AF10* fusion
	t(11;17)(q23;q21)	*MLL/AF17* fusion
M7	inv(8)(p11q13)	*MOZ/TIF2* fusion
AML general	inv(3)(q21q26) and t(3;3)(q21;q26)	*EVI1* dysregulation
	t(3;5)(q25.1;q34)	*NPM/MLF1* fusion
	inv(11)(p15q22)	*NUP98/DDX10* fusion
	t(11;16)(q23;p13)	*MLL/CBFβ* fusion
	t(11;22)(q23;q13)	*MLL/p300* fusion
	+11, 11q+	Partial tandem duplication *MLL*

Hematological findings and symptoms of AML

The hematological findings and symptoms of this disease group include:

- normochromic, normocytic anemia;
- the white cell count can be decreased, normal or increased;
- blast cells in blood films (varying number), Auer rods may be present, or promyelocytes, agranular neutrophils and myelomonocytic cells, depending on the maturation stage of the abnormal cells in disease;
- pallor, lethargy, anemia, fever, malaise, features of infections, septicemia;
- spontaneous bruising, purpura, bleeding gums;
- hypercellular bone marrow, 75% of total marrow cells are usually blast cells;
- thrombocytopenia;
- disseminated intravascular coagulation (DIC) in AML M3;
- tender bones (in children particularly);
- gum hypertrophy and infiltration, rectal ulceration, skin involvement.

2.2   Myelodysplastic syndromes (MDS) and myeloproliferative disorders (MPD)

Classification of MDS

Five categories of MDS are recognized using guidelines set down by the FAB group (Bennett *et al.*, 1982, 1985; Bain, 1990) and cytogenetics and immunophenotyping may also be incorporated as proposed by the MIC group (Second MIC Cooperative Study Group, 1988). Features of these various groups are shown in *Table 7.3*.

Hematological findings and symptoms of MDS

The hematological findings and symptoms of this disease group include:

- a wide range of blood and bone marrow abnormalities, macrocytosis, ringed sideroblasts, megaloblastic erythropoiesis, disordered granulopoiesis and abnormal megakaryocytes;

Table 7.3 FAB classification of MDS excluding CMML

Disease	Incidence in MDS (%)	% blasts	Other features
Refractory anemia (RA)	30–40	<5 in BM ≤1 in PB	<15% ringed sideroblasts in erythroblasts
Refractory anemia with sideroblasts (RAS)	15–25	<5 in BM <1 in PB	>15% ringed sideroblasts in erythroblasts
Refractory anemia with excess blasts (RAEB)	15–25	5–20 in BM	
Refractory anemia in transformation (RAEBT)	15–20	21–29[a] in BM >5 in PB	Also RAEB if Auer rods present irrespective of blast count

[a]A blast count of >30% is a diagnostic criterion for AML.
BM = bone marrow; PB = peripheral blood.

- qualitative and quantitative abnormalities in one or more of the three myeloid lines: red cells, granulocytes and monocytes, and platelets;
- anemia; and
- infections due to impaired phagocytic production and/or function.

Classification of MPD

Myeloproliferative disorders (MPD) also include the CML and CGL groups. *Table 7.4* shows the features of these categories.

Hematological findings and symptoms of MPD

Chronic granulocytic leukemia and chronic myeloid leukemia

- leukocytosis is usually >50×10^9 to 500×10^9 l^{-1};
- complete spectrum of myeloid cells in blood levels of neutrophils and myelocytes exceeds the blast cells and promyelocytes;
- hypercellular marrow with granulopoietic predominance;
- weight loss, lassitude, anorexia and night sweats;
- splenomegaly nearly always present and sometimes massive causing discomfort, pain or indigestion.

Chronic myelomonocytic leukemia

- anemia;
- reticulocytopenia;
- hypercellular marrow;
- erythroid hyperplasia;
- dyserythropoiesis;
- monocytosis in blood;
- increased marrow monocyte precursors.

Polycythemia rubra vera

- Hemoglobin, hematocrit and red cell count increased;
- neutrophil leukocytosis (half of cases);

Table 7.4 Classification of myeloproliferative disorders

Disease	Major proliferative component	% with chromosome abnormality
Chronic myeloid leukemia (CML) and chronic granulocytic leukemia (CGL)	Myeloid activity predominates granulocytic and megakaryocytic hyperplasia, high white cell count	90
Chronic myelomonocytic leukemia (CMML)	Myeloid activity predominates granulocytic and megakaryocytic hyperplasia, high white cell count	30
Polycythemia rubra vera (PRV)	Red cell activity predominates, erythrocytosis	15
Essential thrombocythemia (ET)	Platelet activity predominates, thromboembolitic and hemorrhagic phenomena	5
Myelofibrosis (MF) myeloid metaplasia	Reactive marrow fibrosis predominates	50

- raised platelet count (half of cases);
- neutrophil alkaline phosphase score increased, raised serum vitamin B_{12} binding capacity;
- headaches, pruritis, dyspnea, blurred vision and night sweats;
- retinal venous engorgement, conjunctival suffusion;
- splenomegaly (two-thirds of cases);
- hemorrhage or thrombosis;
- gout.

Essential thrombocythemia

- abnormal large platelets and megakaryocyte fragments in blood film;
- platelet count raised above 1000;
- platelet function tests abnormal;
- anemia;
- massive splenomegaly, giving discomfort, pain or indigestion;
- weight loss, anorexia and night sweats;
- bleeding problems, bone pain.

Myelofibrosis, myeloid metaplasia

- splenomegaly;
- anemia;
- normochromic, normocytic red blood cells;
- weight loss;
- platelets may be high.

3.   Chromosomal abnormalities observed in acute myeloid leukemia

3.1   Acute myeloid leukemia – FAB category M0

This is a difficult group to classify, as these cases of AML do not fit into any of the FAB criteria. The blasts are agranular, have no Auer rods, and are myeloperoxidase negative. They are also morphologically similar to ALL L2. However, the anti-myeloperoxidase antibody used in immunophenotyping is useful for the diagnosis of this poorly differentiated leukemia.

Chromosome changes observed in this group are listed below:

- deletion of chromosomes 5q and 7q;
- monosomy of chromosomes 5 and 7;
- trisomy 13.

3.2   Acute myeloid leukemia – FAB category M1

This group of poorly differentiated leukemias is again a difficult group to classify on morphology alone, and immunophenotyping is again useful in the diagnosis and classification.

Cytogenetically, there is no particular abnormality associated with AML M1, and often these cases are frustrating, as chromosome abnormalities are either not present or perhaps undetectable. However, certain changes are seen in association with this group, and they are as follows:

- monosomy of chromosomes 5 and 7;
- deletion of chromosomes 5q, 7q and 9q;
- trisomy of chromosomes 8, 13, 11 and 21;
- abnormalities of chromosome 3q;
- translocation t(9;22)(q34;q11).

3.3 Acute myeloid leukemia – FAB category M2

Translocation t(8;21)(q22;q22) (see *Figure 7.2*) is usually associated with AML M2 with eosinophilia and the blast cells normally contain a single Auer rod, but can also be observed in AML M1 and M4. Secondary changes include the loss of a sex chromosome, trisomy 8, del(9q) and monosomy 7. Del(9q), trisomy 8 and monosomy 7 can all be observed as sole abnormalities. This change is observed in younger AML patients with a reasonable prognosis. It is observed in about 8% of all AML cases.

Other changes observed in AML M2 are:

- trisomy of chromosomes 4, 8, 11 and 21;
- monosomy of chromosomes 5 and 7;
- deletion of chromosomes 5q and 7q;
- translocation t(2;4)(p23;q35);

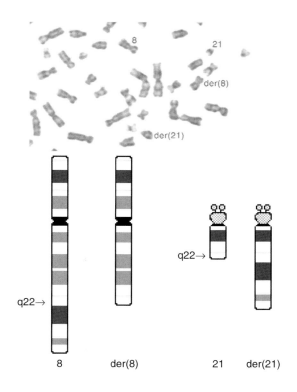

Figure 7.2

Translocation t(8;21) showing a partial metaphase and a diagrammatic representation.

- translocation t(7;11)(p15;p15);
- translocation t(11;20)(p15;q11).

3.4 Acute myeloid leukemia – FAB category M2-baso

Translocation t(6;9)(p23;q34) is a rare and subtle rearrangement, which results in the fusion of *DEK* on chromosome 6 and *CAN* on chromosome 9. It is usually observed in younger patients, with multilineage disease. Occasionally a history of toxic exposure is noted (see *Figure 7.3*).

Deletion of chromosome 12p involving bands 12p11–12p13 is also observed in this category. This abnormality may occur in *de novo* AML or in secondary leukemia. The larger deletion [i.e. del(12)(p11p13)] appears to have a worse prognosis than the more subtle deletions.

3.5 Acute promyelocytic leukemia – FAB category M3 and M3V

Translocation t(15;17)(q22;q21) (*Figure 7.4*) is a highly specific abnormality which is *only* observed in AML M3 and AML M3V. In the past these patients had a poor prognosis, as conventional AML therapy was utilized for patients with this abnormality. However, the prognosis is now greatly improved by treatment with all-*trans* retinoic acid (ATRA). The bone marrow can be problematic for the cytogeneticist as the DIC, a common symptom the patient suffers, causes rapid clotting of the aspirated marrow and it can occasionally be difficult to produce preparations that yield sufficient reasonable-quality metaphases. It is helpful to employ the FISH probe for the *PML/RARα* fusion to study interphase cells if no metaphases

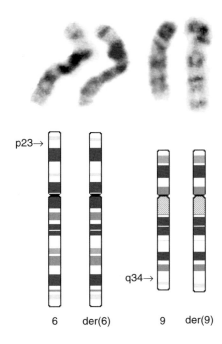

Figure 7.3

Translocation t(6;9) showing a partial karyotype and a diagrammatic representation.

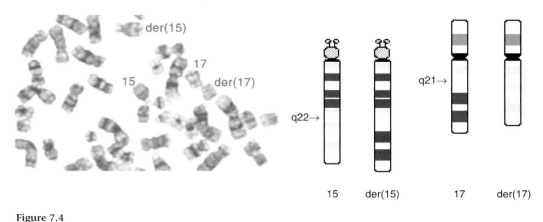

Figure 7.4

Translocation t(15;17) showing a partial metaphase and a diagrammatic representation.

were present in the original preparation. Morphologically, it is difficult to distinguish M3 from M3V, and in this instance cytogenetic findings are crucial to confirm the presence or absence of the t(15;17). Additional changes observed are +8, +21 and ider(17).

The t(11;17)(q23;q12) variant of the t(15;17) translocation does not show the same response to ATRA, although M3 morphology is apparent. There are some reports suggesting that an increased dose of ATRA in these patients may have greater success.

3.6   Acute myeloid leukemia – FAB category M4Eo

Inversion inv(16)(p13q22) (*Figure 7.5a*) and translocation t(16;16)(p13;q22) (*Figure 7.5b*) are associated with AML M4Eo. The rearrangement of 16q22 has been reported in approximately 4% of all AML cases. Nearly all are pericentric inversions, resulting in the *CBFβ/MYH11* fusion. This change is thought to be associated with a reasonable prognosis. Secondary changes include +8 and del(7q). Trisomy 22 is also a secondary change; although this can be observed as a sole abnormality, it is rare elsewhere. The presence of trisomy 22 should alert the cytogeneticist to the possibility of an inversion of chromosome 16 in the sample. The inverted 16 can also be observed in a range of FAB types and occasionally in MDS. However, in MDS the abnormality of chromosome 16 is usually a deletion.

3.7   Acute myeloid leukemia – FAB categories M4 and M5

Both the myelomonocytic and monocytic leukemias have an association with abnormalities of 11q23 (*Figure 7.6*). Younger patients, particularly infants with leukemia, secondary leukemia and AML FAB types M4 and M5 usually exhibit abnormalities of 11q23. Biphenotypic and minimally differentiated leukemia are also known to show 11q changes. The region 11q23 is a 'hot spot' for disease, and is involved in a wide range of hematological disorders. Disruption of the *MLL* (*HRX, ALL1* or *HTRX*) gene occurs via a number of mechanisms depending upon the disease type being studied. A translocation producing the novel *MLL* fusion gene is the main

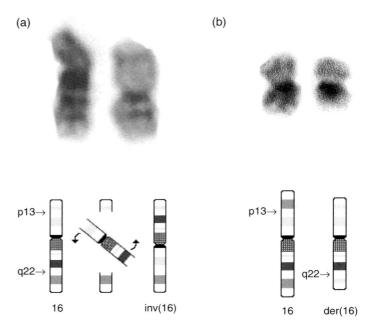

Figure 7.5

(a) Inversion of a chromosome 16 inv(16), and (b) translocation t(16;16), showing partial karyotypes and diagrammatic representations.

mechanism involved, and many chromosomes have been involved in these translocations. The most frequent of these are shown below:

- t(4;11)(q21;q23) – some myeloid cases mainly M4 but more frequent in ALL;
- t(9;11)(p21;q23) – mostly myeloid cases, M5a;
- t(11;19)(q23;p13.1 or p13.3) – cases with 19p13.1 breakpoint, all myeloid, M4 or M5a; some cases with 19p13.3 breakpoint, M4 or M5a;
- t(6;11)(q27;q23) – mostly myeloid cases, M4 or M5a;
- t(10;11)(p12;q23) – mostly myeloid cases, M5a.

Many less common partner sites have been studied and have been reviewed by the European Workshop (Secker-Walker, 1998). These include 1p32, 1q21, 2p21, 17q12–21 and 17q25.

Trisomy 4 is also observed in association with AML M4 but can be observed in M1, M2 or M0 and usually carries a poor prognosis. Double minutes are a feature sometimes associated with this change (see *Figure 4.6* in *Chapter 4*).

3.8   Acute myeloid leukemia – FAB category M5a

Translocation t(8;16)(p11;p13) is found associated with acute leukemia with monocytic differentiation. Such cases often show hemophagocytosis by leukemic cells. Other abnormalities associated with AML M5a include:

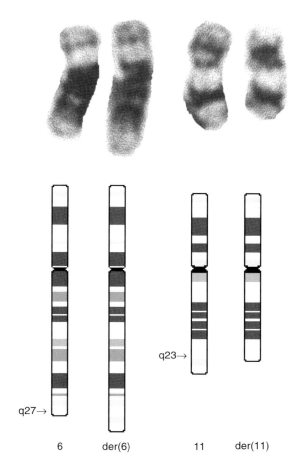

Figure 7.6

Translocation t(6;11) showing a partial karyotype and a diagrammatic representation.

- t(6;11)(q27;q23);
- t(9;11)(p21;q23);
- t(1;12)(p36;p12);
- abnormalities of 11q23.

3.9 Acute myeloid leukemia – FAB category M6

A diagnosis of AML M6 is considered when erythroblasts exceed more than 50% of nucleated cells in the bone marrow, and MDS can be excluded if blasts exceed 30% of nonerythroid cells. There are no specific chromosomal changes observed in this rare group of diseases, but some have been associated with AML M6 and are shown below.

A number of changes involving the breakpoints on 3q21 and 3q26 are associated with megakaryocytic abnormalities and trilineage dysplasia. All appear to have a poor prognosis.

Inversion of chromosome 3, inv(3)(q21q26) and translocation t(3;3)(q21;q26) (*Figure 7.7*) are observed in <1% of AML (in a range of FAB types), MDS and also CGL. Monosomy 7 is often observed as a secondary change. These abnormalities show a dysregulation of the gene *EV11*.

Other changes include:

■   monosomy of chromosomes 5 and 7;
■   trisomy 8;
■   deletion of chromosomes 5q, 7q, 9q and 20q;
■   isochromosome for 21q.

3.10   Acute myeloid leukemia – FAB category M7

This group is characterized by a predominant megakaryoblastic component. Chromosome changes observed in this group include:

■   Translocation t(1;22)(p13;q13) – occurs in infants and young children with hepatosplenomegaly;
■   hyperdiploidy is common with extra derivative chromosome 1, trisomy for chromosomes 6, 17, 19 and 21;
■   inversion or deletion of chromosome 3;
■   trisomy of chromosomes 8 and 21;
■   deletion of chromosome 20q;
■   ring chromosomes.

4.   Chromosome changes in myelodysplastic syndromes

Although there are various categories of MDS as described above, the chromosome changes observed in MDS are usually spread across all the categories and do not normally pinpoint any particular group. However, there are abnormalities associated with MDS in general, and specific syndromes such as the two described below deserve attention.

The 5q- syndrome   Although this is a common change in myeloid disorders when observed as the sole abnormality in MDS, it usually indicates a subset of MDS patients with benign dysplasia, who are usually elderly, female, with refractory macrocytic anemia and abnormalities of megakaryocytes. The risk of transformation to AML is low in these types of patients.

Deletion of chromosome 5q is also often observed as part of a complex karyotype also involving chromosomes 3, 7, 12 and 17. This is particularly common in secondary leukemias. When a patient presents with symptoms indicating MDS transforming to AML, these type of complex karyotypes are particularly common, involving both chromosomes 5 and 7. Prognosis is usually poor in those patients with a complex karyotype.

Monosomy 7 syndrome   This abnormality is associated with secondary disease or therapy-related MDS. Monosomy 7 is recognized as a recurring abnormality in MDS cases presenting in childhood. Patients develop or present with severe thrombocytopenia, and transformation to AML is common, the prognosis usually being very poor.

Other chromosomal changes in MDS can also be observed in AML and are shown below:

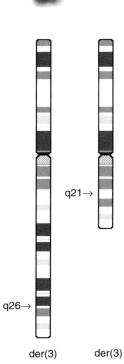

q21→

q26→

der(3)   der(3)

Figure 7.7

Translocation t(3;3) showing a partial karyotype and a diagrammatic representation.

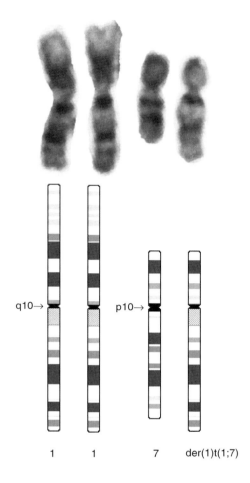

q10→   p10→

1   1   7   der(1)t(1;7)

Figure 7.8

Unbalanced translocation t(1;7) showing a partial karyotype and a diagrammatic representation. Note that the p arm of chromosome 1 and the q arm of chromosome 7 are lost.

- the unbalanced structural change +1,der(1;7)(q10;p10), which results from the fusion of the q arm of chromosome 1 and the p arm of chromosome 7 giving rise to trisomy for 1q and monosomy for 7q (see *Figure 7.8*); the prognosis is poor; this abnormality is thought to be associated with therapy-related disease;
- monosomy 5 is observed in MDS but can also be seen in MPD and AML, and secondary disease;
- trisomy 6 is a rare observation, although it has been noted in hypoplastic MDS;
- deletion of 7q is also associated with abnormalities of chromosome 5q in secondary disease. Breakpoints are variable with the most commonly reported sites being 7q22 and 7q32-7qter. Molecular studies have demonstrated that some of these deletions prove to be derivative chromosome 7s;

■ trisomy 8 is the most common abnormality observed in myeloid disease, found in about 9% of AML and in 4% as the sole change in addition to its association with MDS; it is very unusual to find this change in lymphoid malignancy;

■ deletion of 11q, as mentioned previously, also associated with AML;

■ deletion of 12p, often also found in secondary disease and usually associated with a poor prognosis;

■ trisomy 14 is usually associated with MDS;

■ trisomy 15 is occasionally observed in MDS but has also been seen in cases without confirmed hematological disease;

■ isochromosome for 17q is normally associated with the translocation t(9;22)(q34;q11) (see *Section 5.1*), but can be observed as sole abnormality, where a poor prognosis is normally indicated. Deletions of the p arm of chromosome 17 do occur, as do ring chromosome 17s;

■ trisomy 21 is found in MDS and AML of no particular FAB type and is often observed in clonal evolution;

■ loss of chromosome Y is considered to be an age-related phenomenon of no known clinical significance, if observed alone. Loss of the Y chromosome is observed with t(8;21) and has no impact upon prognosis if observed as such a secondary change; loss of the X chromosome is very rare, but again can be observed in association with t(8;21).

5.   Chromosome changes in myeloproliferative disorders

5.1   Chronic myeloid leukemia

Chronic myeloid leukemia is also confusingly known as chronic myeloblastic leukemia and chronic granulocytic leukemia. Some people use the term CGL for cases which have the associated translocation t(9;22), giving rise to the Philadelphia chromosome (Ph chromosome), and use CML to include atypical forms of the disorder, chronic myelomonocytic leukemia (CMMoL), eosinophilic leukemia, chronic neutrophilic leukemia (CNL) and also juvenile CML (jCML). This author prefers the use of the term CML as the Philadelphia chromosome is not restricted to cells of granulocytic lineage. As already mentioned, the abnormality most commonly associated with CML is translocation t(9;22)(q34;q11) (*Figure 7.9*). This group of diseases has a vastly variable benign chronic phase (mean duration of approximately 3 years). Transformation to the acute phase is either a gradual clinical acceleration or a very rapid blast crisis, the latter being less common. More than 30% of blasts should be present in the peripheral blood or bone marrow to categorize the disease as acute phase.

In the chronic phase, the Philadelphia chromosome formed from the t(9;22) translocation is observed in most cases. Variant translocations also occur where a third chromosome is involved. These groups comprise about 95% of cases. In the remaining 5%, half of the cases show a molecular rearrangement of the *ABL* gene on chromosome 9 and the *BCR* gene on chromosome 22. It is thought that the other half of this 5% can be classified as a different disease type altogether, with no visible Philadelphia

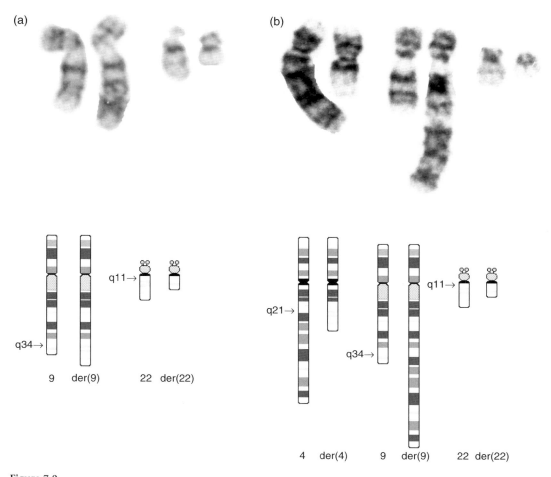

Figure 7.9

(a) Translocation t(9;22) showing a partial karyotype and a diagrammatic representation; and (b) a complex three-way translocation t(4;9;22) showing a partial karyotype and a diagrammatic representation.

chromosome translocation, and no apparent molecular rearrangement of BCR/ABL.

Translocation t(9;22), however, is not exclusive to CML as it can occasionally be observed in *de novo* AML and also in ALL (see *Section 6.5*).

Additional chromosome changes are observed in the accelerated phase of CML and can also be observed even when the clinical symptoms do not give any indication of such a change. This is important for the clinician, as any extra changes should be treated with suspicion. The most common changes are shown below and these can occur alone or together in a variety of combinations:

- trisomy 8 (8%);
- i(17)(q10) and 17p abnormalities (12%);
- +der(22)t(9;22) extra Ph chromosome (14%);
- trisomy 19 (1%);

■ trisomy 8 and i(17q) (9%);
■ trisomy 8 and extra Ph chromosome (9%);
■ trisomy 8, trisomy 19 and extra Ph chromosome (7%);
■ trisomy 8, i(17q) and extra Ph chromosome (3%);
■ trisomy 8, trisomy 19, i(17q) and extra Ph chromosome (2%);
■ trisomy 19 and extra Ph chromosome (3%);
■ translocation t(3;21)(q26;q22) (<1%).

Identification of the blast type is usually by immunophenotyping, as extra cytogenetic changes are not normally confined to a particular type of acute disease. For example, the isochromosome 17q is generally associated with myeloid transformation, but it is still not possible to confirm a myeloid transformation by cytogenetics alone. Trisomy 8 and the extra Philadelphia chromosome are also observed in the chronic phase without any indication of transformation to acute leukemia.

5.2 Polycythemia rubra vera

The deletion of chromosome 20 in the q arm is a common abnormality in myeloid disorders, but is most frequently associated with MPD, particularly polycythemia rubra vera (PRV). Patients with either the del(20)(q11q13.1) or del(20)(q11q13.3) show features of dysplasia in erythroid precursors and megakaryocytes (*Figure 7.10*). It occurs in 5% of MDS patients and also in AML, particularly AML M6.

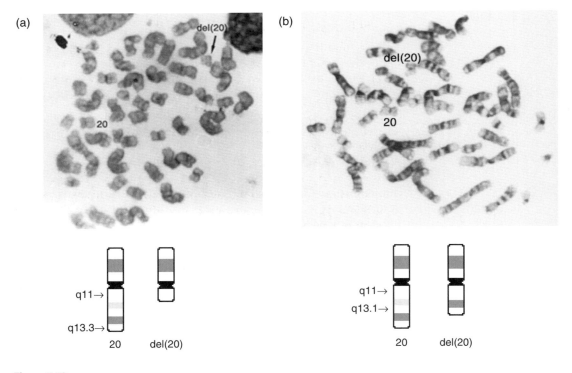

Figure 7.10

Deletion of chromosome 20 showing metaphase spreads with the abnormal and normal chromosome 20s indicated, and a diagrammatic representation of both (a) the larger deletion and (b) the smaller deletion.

5.3 Chromosomal changes in other myeloproliferative disorders

■ Deletion of chromosome 13q is observed in myelofibrosis and MPD, but is more commonly observed in chronic lymphoid leukemia (see *Section 7*).

■ Trisomy for 1q is observed in hematological disorders as well as in solid tumors. In myeloid disorders it is most often observed in MPD.

■ Trisomy 9 is typically associated with MPD.

6. Acute lymphoblastic leukemia

6.1 Classification of acute lymphoblastic leukemias

Morphological evaluation has been the accepted method for classification of the various cytological types of acute and chronic leukemia for over a hundred years. Cytochemical stains and, in recent years, immunocytochemistry aid the distinction of lymphoid and myeloid leukemias, characterizing leukemic blast cells.

Advances in cytogenetics (high-resolution banding and molecular probes) have helped to provide further characterization of the subgroups of acute lymphoblastic leukemia (ALL) by identifying consistent chromosomal changes that may not have been apparent on initial G-banding examination [such as the t(12;21) translocation, *Section 6.5*]. The understanding of the various cytogenetic changes that occur in ALL is enhanced by awareness of the morphological characteristics and their immunological features.

6.2 FAB classification system

The classification of ALL by the FAB cooperative group (First MIC Cooperative Study Group, 1986) has been used as a means of distinguishing the three main subtypes of ALL, shown below.

■ ALL FAB type LI – the blast cells are small and uniform, with a high nuclear to cytoplasmic ratio. The nuclei are mainly round, with inconspicuous nucleoli. The cytoplasm is scanty, and slightly to moderately basophilic.

■ ALL FAB type L2 – the blast cells in L2 are larger than the blasts in L1, and, in addition, more heterogeneous in appearance. The nuclear to cytoplasmic ratio is lower than that of L1. Because of this heterogeneity, this FAB type sometimes appears as an undifferentiated leukemia. The nuclear outline tends to be indented or irregular in a number of cells. The amount of cytoplasm varies, as does cytoplasmic basophilia.

■ ALL FAB type L3 – the blast cells tend to be larger than L1 or L2 and are homogeneous, with strongly basophilic cytoplasm. Most cells show a vacuolation of the cytoplasm. Although the nuclear to cytoplasmic ratio is high, it is not as high as noted in L1. The nuclei tend to be mainly round in shape.

6.3 Immunophenotyping

It should be noted that the monoclonal antibodies used to define cell surface antigens on blasts identify both the normal and leukemic blasts. The applications of immunophenotyping for hematological disorders can be summarized below:

- separating ALL from AML, particularly in cases where difficult interpretation problems arise from morphology and cytochemistry results;
- defining B cell precursor ALL from B cell ALL and T cell ALL;
- identifying certain subtypes of AML;
- detection of CD41a, CD42b and CD61, which diagnose AML M7;
- distinguishing between B and T cell chronic lymphoproliferative disorders.

The four major immunological classes of B lineage ALL, which comprise the majority of cases, are listed below and all show common immunological features of DR+, Tdt+, CD19+ and CD34+:

- early B precursor ALL (previously Null-ALL);
- pre-pre-B ALL or pro-B ALL (previously common ALL);
- pre-B ALL;
- B cell ALL.

The three classes of T cell ALL, which represent 15% of ALL cases, are listed below:

- pre-T or prothymocyte T ALL;
- intermediate or immature T ALL;
- mature thymocyte T ALL.

The T ALL cases are almost equally distributed between the above three categories, and within the mature thymocyte T ALL group, 50% of cases show a rearrangement of α and β T cell receptor genes, and the other 50% show a rearrangement of the γ and δ T cell receptor genes.

Lymphocytes are the major cellular elements of the body's immune system. B lineage lymphocytes are responsible for immunoglobulin (Ig)-mediated immunity and the T lineage lymphocytes play a role in cellular immunity. When B lymphocytes mature, somatic recombination of the Ig loci occurs prior to Ig production. These Ig loci are located at the following chromosomal bands:

- 14q32 Ig heavy chain gene;
- 2p12 κ light chain gene;
- 22q11 λ light chain gene.

The three loci shown below with their chromosomal location are essential in coding for T cell receptor components vital for T cell differentiation:

- 14q11.2 α chain locus;
- 7q35 β chain locus;
- 7p15 γ chain locus;
- 14q11 δ chain locus.

6.4 Hematological findings and symptoms of disease

The hematological findings and symptoms of ALL include:

- normochromic, normocytic anemia;
- the white cell count can be decreased, normal or increased;
- blast cells in blood films (varying number);

- hypercellular bone marrow; typically 75% of total marrow cells are blast cells;
- pallor, lethargy, anemia, fever, malaise, features of infections, septicemia;
- spontaneous bruising, purpura, bleeding gums;
- tender bones (in children particularly);
- superficial lymphadenopathy, moderate splenomegaly, hepatomegaly;
- meningeal syndrome, headache, nausea and vomiting;
- testicular swelling;
- mediastinal compression.

6.5   Chromosome abnormalities observed in ALL

Early B precursor and pre-pre-B cell ALL

The translocation t(9;22)(q34;q11) (see *Section 5.1*) results in the Philadelphia chromosome, the der(22)t(9;22)(q34;q11). Although diagnostic for CML, this translocation is present in 25–30% of adult ALL and approximately 5% of childhood ALL, and most often observed with an ALL L1 or L2 FAB type. Immunophenotyping is usually of pre-pre B cell type (i.e. CD10 positive and cytoplasmic immunoglobulin negative), but some cases can be early B precursor or pre-B phenotype. A normal cell population is present in most cases of ALL and this normally distinguishes ALL cases from those of CML. The Philadelphia chromosome translocations also occur in hyperdiploid karyotypes of over 50 chromosomes. As the t(9;22) is a poor prognostic sign, whereas the hyperdiploid karyotype is considered to carry a better prognosis and a better response to conventional therapy, full analysis of any hyperdiploid karyotype is essential.

The molecular consequence of the t(9;22)(q34;q11) is the fusion of *ABL* at 9q34 with *BCR* at 22q11, forming a chimeric *BCR/ABL* gene which produces a chimeric protein with increased tyrosine kinase activity. Breakpoints at *BCR* cluster in two regions: the major breakpoint cluster region (*M-BCR*), which produces a 210 kDa protein; and the minor breakpoint cluster region (*m-BCR*) which produces a 190 kDa protein. Both products can be observed in ALL, although the 190 kDa protein predominates, which differs from the situation in CML. This provides a clinical dilemma as to whether Ph+ve ALL patients expressing the 210 kDa protein have a form of lymphoid acute phase of CML or a *de novo* acute leukemia.

Translocation t(4;11)(q21;q23) (see *Figure 7.11*) is observed in 40–50% of infants with ALL. It is more common in females. Patients usually present with hepato-splenomegaly, CNS involvement and extremely high white cell counts with a high percentage of blasts. The leukemic cells are usually very immature blasts of early B precursor or pro-B type and of L2 FAB type. Some cases appear to be biphenotypic with monocytic or myelomonocytic characteristics. Prognosis is very poor, particularly for infants and adult patients. The t(4;11) translocation gives rise to the *MLL/AF4* fusion.

In contrast to t(4;11), translocation t(11;19)(q23;p13.3) involves a subtle exchange of only small chromosomal segments and therefore detecting the translocation in poor chromosome preparations becomes difficult. The clinical and immunophenotypic features of the disease are similar to that observed with the t(4;11), except that biphenotypic leukemia

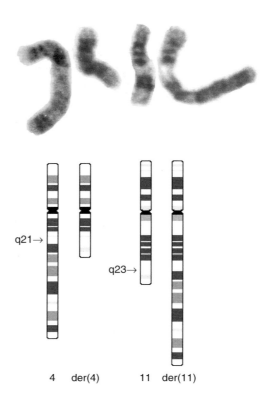

Figure 7.11

Translocation t(4;11) showing a partial karyotype and a diagrammatic representation.

and AML are more common in these instances. This translocation is observed in approximately 10% of infants with ALL. Prognosis is poor for infants and adults. The t(11;19) translocation gives rise to the *MLL/ENL* fusion. The *MLL* gene is often involved in infant or congenital leukemia, and also associated with leukemias that demonstrate myelomonocytic or monocytic characteristics.

Dicentric dic(9;12)(p11–13;p11–12) is observed in approximately 1% of ALL patients, almost exclusively in early B lineage leukemia, and is most common in older children and young adults. Remission rates are high and prognosis is good. Trisomy 8 is often seen as a secondary change with this abnormality. It is usually observed in the ALL L1 category.

Dicentric dic(9;20)(p11–13;q11) is a subtle chromosome abnormality, observed mainly in early B lineage ALL L1, and it often masquerades as monosomy 20. Trisomy 21 is also a frequent secondary finding. Remission rates are good for patients with this change.

Deletion of chromosome 6q occurs in about 4–6% of childhood ALL and less frequently in adults. It can be observed either as the sole change or in association with other abnormalities. It is a common observation in lymphoid disease. The breakpoints are variable but usually occur between q14 and q21. These cases are usually CD10 (c-ALLA)-positive, B cell ALL L2, but can be observed in T cell disease.

Hyperdiploidy with greater than 50 chromosomes is observed in 20–30% of childhood ALL and approximately 5% of adult ALL. Patients with >50 chromosomes generally have clones of between 51 and 68 chromosomes. High hyperdiploidy is associated with female patients, low white cell count, having FAB type L1 or L2 and a pre-pre-B ALL immunophenotype. Prognosis is good with long term survival of 70–80% in children. A typical pattern of chromosome gain emerges showing extra copies of chromosomes 4, 6, 10, 14, 18, 21 and X. Nonrandom translocations are occasionally present, including t(9;22), t(4;11), t(1;19) and t(12;21), with the translocation probably being the primary change. In such cases the leukemia should be classified according to the translocation rather than the ploidy group in order to categorize the patient into the correct prognostic group (*Figure 7.12*).

Translocation t(12;21)(p13;q22) had not previously been detected by conventional banding techniques, and has only been disclosed recently by molecular techniques. It has now been observed in 20–30% of B lineage childhood ALL, being the most frequent nonrandom translocation in this ALL subtype. The immunophenotype pattern of this abnormality can be early B precursor pre-pre-B or pre-B ALL. In suspecting the t(12;21) it should be noted that a deletion of chromosome 12p is often observed as a secondary change. An apparent extra chromosome 21 or isochromosome 21q being present should also alert a cytogeneticist to the possibility of this subtle change as these may actually be the der(21)t(12;21) or the isoderivative chromosome 21. One must always utilize FISH or RT-PCR

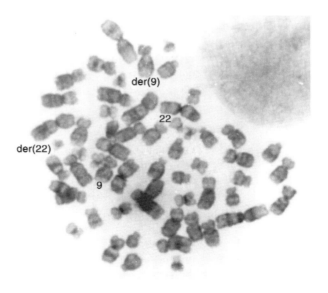

Figure 7.12

Hyperdiploid metaphase spread showing a t(9;22) translocation.

techniques in suspicious cases or even those where no abnormalities are apparent. The t(12;21)(p13;q22) involves a fusion of the *TEL* gene on 12p13 to the *AML1* gene on chromosome 21q22, giving rise to a *TEL/AML1* fusion protein. The *TEL/AML1* rearrangement carries a very good prognosis in childhood B cell ALL.

Pre-B cell ALL

The t(1;19)(q23;p13) translocation, and its rarer variant t(17;19)(q21–22;p13), are observed in about 25% of pre-B cell ALL patients. The cells usually have ALL L1 morphology. The t(1;19) translocation is normally observed in two forms:

- a balanced translocation t(1;19) (25% cases);
- an unbalanced translocation, showing der(19)t(1;19) and two normal homologs of chromosome 1 (75% of cases).

The inexperienced cytogeneticist may be unaware of the appearance of the derivative chromosome 19, and in poor cells interpretation may be difficult (see *Figure 7.13*). This reflects the importance of being alert to the appearance of translocation products.

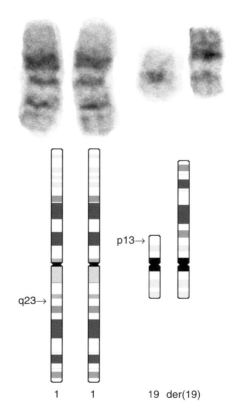

Figure 7.13

Unbalanced translocation t(1;19) showing a partial karyotype and a diagrammatic representation.

Mature B cell ALL L3

The translocations t(8;14)(q24;q32), t(2;8)(p12;q24), and t(8;22)(q24;q11) are observed most frequently in mature B cell ALL and Burkitt's lymphoma. The translocations are almost always associated with ALL L3 morphology. The t(8;14) is the most frequent of the variants (see *Figure 7.14*).

The patients are usually male, with a median age of 11 years in children and 40 years in adults. Bulky extramedullary disease is apparent upon presentation and patients frequently have central nervous system (CNS) involvement, an unusually high blast cell count and increased proliferative rate, and inevitably poor prognosis. However, the introduction of certain short-term dose-intensive regimens has improved the clinical courses of these patients.

The common feature in these translocations is the break in the oncogene *MYC*, located at band 8q24. The regulation of this oncogene is abnormal following the translocation, the crucial event being the juxtapositioning of *MYC* with the locus of the immunoglobulin heavy-chain gene at 14q32. In the variant translocations, *MYC* is brought into juxtaposition with the immunoglobulin kappa (κ) light chain gene locus (at 2p12), in the t(2;8), or the immunoglobulin lambda (λ) light chain gene locus (at 22q11), in the t(8;22). The rearrangements bring *MYC* under the influence of transcription-sequence regulating sequences of the active immunoglobulin locus, resulting in dysregulation of *MYC*, increased transcription, and consequently neoplastic growth.

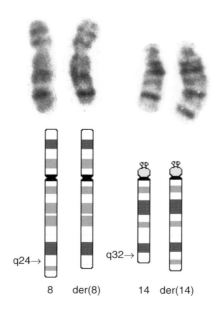

8   der(8)   14   der(14)

Figure 7.14

Translocation t(8;14) showing a partial karyotype and a diagrammatic representation.

T lineage ALL

Both the clinical features and chromosome abnormalities seen in T lineage ALL show similarities with those observed in the T cell chronic lymphoproliferative disorders. This is more common in males than females and usually affects older children and young adults. Symptoms include a high white cell count, lymphadenopathy, splenomegaly, mediastinal mass, and CNS involvement. Prognosis is poor.

The translocations observed often involve one of the T cell receptor (*TCR*) genes located at bands 14q11.2 (*TCRα/δ*), 7q35 (*TCRβ*) and 7p14–15 (*TCRγ*). Oncogene activation occurs in a similar fashion to that observed in B cell ALL. The translocations shown below are extremely rare:

- t(1;7)(p32;q35) – deregulating *TAL1* by proximity to *TCRβ* at 7q35;
- t(1;14)(p32-34;q11) – deregulating *TAL1* by proximity to *TCRδ* at 14q11;
- t(7;9)(q34;q32) – deregulating *TAL2* by proximity to *TCRβ* at 7q35;
- t(7;10)(q34–35;q24) – deregulating *HOX11* by proximity to *TCRβ* at 7q35;
- t(7;11)(q35;p13) – deregulating *RBTN2* by proximity to *TCRβ* at 7q35;
- t(7;19)(q34–35;p13) – deregulating *LYL1* by proximity to *TCRβ* at 7q35;
- t(8;14)(q24;q11) – deregulating *MYC* by proximity to immunoglobulin heavy chain gene at 14q32;
- t(10;14)(q24;q11) – deregulating *HOX11* by proximity to *TCRδ* at 14q11;
- t(11;14)(p15;q11) – deregulating *RBTN1* possibly by proximity to *TCRδ* at 14q11;
- t(11;14)(p13;q11) – deregulating *RBTN2* possibly by proximity to *TCRδ* at 14q11.

Abnormalities of 9p, including deletions and unbalanced translocations, are present in approximately 10% of ALL, mainly in T lineage ALL, but they have also been observed in B lineage ALL.

Inversion of chromosome 14 inv(14)(q11q32) and translocation t(14;14)(q11;q32) have been observed in T cell disease, associated with adult T cell leukemia/lymphoma (ATL/L), but have also been observed in T cell CLL. As such, these changes are probably subtypes in T cell malignancies and not specific to ATL.

6.6   Other abnormalities in ALL

Near-haploidy (23–28 chromosomes) is usually classified with <30 chromosomes, normally between 23 and 28 chromosomes. The near-haploid cells can be interpreted as broken metaphase spreads and analysis of the poorest quality cells is essential to deduce whether this is indeed artifact or true near-haploidy. The chromosomal pattern in near haploidy usually shows two copies of chromosomes 6, 8, 10, 14, 18, 21 and the sex chromosomes, the other chromosomes tending to be lost. A normal diploid clone and a cell line with double the near-haploid number of chromosomes, is often seen alongside these types of cells. The latter hyperdiploid cells in these cases are unusual, showing two or four copies of chromosomes, and this should differentiate this type of case from >50 hyperdiploid ALL which usually has three copies of chromosomes. The near-haploidy group is a rare type of childhood ALL, mainly affecting children and teenagers, and is associated with a poor prognosis.

Severe hypodiploidy (23–29 chromosomes) is also rare and sometimes classed with near haploidy as the prognosis is also poor. However, cytogenetically, the two classes form different groups. Patients with severe hypodiploidy are usually adults. The chromosome number ranges from 30–39 and two copies of 1, 4, 5, 6, 8, 9, 10, 11, 18, 19, 21, 22 and the sex chromosomes are present. As above, the hypodiploid clone can often be misinterpreted as broken metaphases.

Other abnormalities that have been noted in ALL are shown below:

- t(1;3)(p34;p21) – early pre-B cell ALL /L1;
- t(1;14)(p32–34;q11) – T cell ALL / L2;
- t(2;14)(p13;q32) – pre-B cell ALL /L1;
- t(5;14)(q31;q32) – pre-B cell ALL/ L1;
- i(6p) – B cell ALL/ L2;
- t(7;7)(p15;q11) – T cell ALL/ L2;
- t(7;10)(q34–35;q24) – T cell ALL/ L2;
- dic(7;9)(p11;p11) – pre-B cell ALL/ L1 or L2;
- t(12;13)(p13;q14) – T or B cell ALL/ L1 or L2
- dic(12;17)(p12–13;q21) – T or B cell ALL/ L1 or L2;
- t(14;22)(q32;q11) – B cell ALL/ L2;
- t(17;19)(q23;p13) – early pre-B cell ALL / L1 or L2.

7. Chronic lymphoproliferative disorders and lymphomas

7.1 Classification of chronic lymphoproliferative disorders and lymphomas

Using cytology and immunophenotyping, this group of disorders has been classified by the FAB group. There are many classifications based on histological features supplemented by immunophenotyping. The most recent is the Revised European–American Lymphoma (REAL) classification, which also includes some cytogenetic information (Harris *et al.*, 1994). It should be remembered that, even using all the optimal methods at the disposal of the clinician, some atypical cases do not fit neatly into any of the categories, which reflects the difficulty in classifying the diseases of the lymphoid system.

7.2 Established chromosome abnormalities in chronic lymphoproliferative disorders

The Second International Working party on chromosomes in chronic lymphoproliferative disorders has demonstrated that certain chromosome changes appear to be relevant in the pathogenesis of disease (Juliusson *et al.*, 1990). Most of the literature surrounds chronic lymphocytic leukemia (CLL) cases, while specific chromosome abnormalities associated with the other disorders are rarer and, as such, associating the true incidence of findings with disease type is difficult. A summary of chronic B and T cell lymphoid leukemias is given below.

B lineage leukemia Chronic lymphocytic leukemia (B-CLL); chronic lymphocytic leukemia, mixed cell type (CLL/PLL); prolymphocytic leukemia (PLL); hairy cell leukemia (HCL); hairy cell leukemia variant (HCL-v); plasma cell leukemia (PCL).

B lineage lymphoma (leukemic phase) Splenic lymphoma with villous lymphocytes (SLVL); follicular lymphoma (FL); mantle cell lymphoma (MCL); Waldenström's macroglobulinemia (lymphoplasmacytic lymphoma).

T lineage leukemia Chronic lymphocytic leukemia/large granular lymphocytosis (T-CLL/LGL); T cell prolymphocytic leukemia (T-PLL); Sézary's syndrome; mycosis fungoides.

T lineage lymphoma (leukemic phase) Adult T cell leukemia/lymphoma (ATL/L); T cell non-Hodgkin's lymphoma (T-NHL); large granular–lymphocytic leukemia (LGL).

The most common changes observed in these disorders are briefly described below.

- *Deletions of chromosome 6q* – these deletions occur in approximately 6% of chronic lymphoid leukemia cases. They are not associated with any particular disease type but are a fairly consistent marker for lymphoid malignancy.
- *Translocation t(6;12)(q15;p13)* – this may be a specific marker for PLL, and other 12p rearrangements have also been described in T cell PLL.
- *Isochromosome 8q* – this abnormality gives rise to apparent trisomy for 8q and is associated with T cell PLL.
- *Trisomy 12* – this is the most common numerical abnormality observed in B cell CLL and, although it occurs in other B cell lymphoproliferative disorders, it is mainly associated with CLL. Patients with trisomy 12 show an unfavorable prognosis compared to those with a normal karyotype. However, having trisomy 12 as the sole abnormality does not significantly shorten survival as compared to normal karyotypes. Those with extra changes in addition to the trisomy 12 show a worse prognosis.
- *Abnormalities of 11q23* – these are most often observed as interstitial deletions (see *Figure 7.15*) involving 11q21–25, and are observed in 20% of CLL cases. Patients with this deletion tend to be younger, but with an advanced clinical disease state, extensive lymphadenopathy and a poor prognosis.
- *Abnormalities of 13q* – these rearrangements usually involve 13q12–14, and deletions have been noted in 10% of CLL cases. However, molecular probes specific for this region show that this percentage is much higher, at 53%. The retinoblastoma gene *RB1* located at 13q14 is deleted in these cases.
- *Abnormalities of 14q at breakpoint q11* – these abnormalities are usually observed as inv(14)(q11q32) (see *Figure 7.16*) and t(14;14)(q11;q32). As seen in the ALL section (*Section 6.3*), the α and δ T cell receptor gene loci are located at 14q11, and *TCL1* is located at 14q32. Variant translocations include t(X;14)(q28;q32), t(8;14)(q24;q11) and

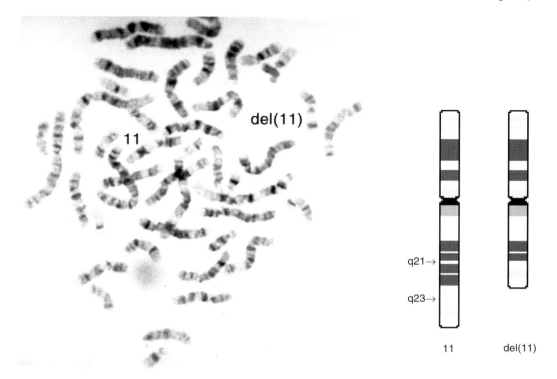

Figure 7.15

Deletion of chromosome 11 in the q arm del(11) showing a metaphase spread with the normal and abnormal chromosome 11s indicated, and a diagrammatic representation.

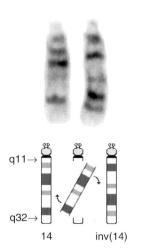

Figure 7.16

Inversion (14)(q11q32), showing a partial karyotype and a diagrammatic representation.

t(11;14)(p13;q11). These changes are associated with LGL and T cell lymphomas and leukemias, including T cell PLL and T cell leukemia.

■ *Abnormalities of 14q at breakpoint q32* – abnormalities at 14q32 are commonly observed in B cell CLL, PLL and HCL. These can also include the t(14;18)(q32;q21) observed in follicle center lymphoma (see *Section 7.4*), although many other translocations involving 14q32 have been described.

■ *t(11;14)(q13;q32)* – this abnormality has been found in B cell CLL, PLL SLVL and PCL involving the *BCL1* gene.

■ *Abnormalities of 17p* – these abnormalities are generally associated with mixed-cell type CLL. Although p53 mutations are common in human cancer in general, cytogenetic changes at 17p are not that frequent in CLL. The p53 mutations are more common in PLL.

7.3   The lymphomas

Lymphomas can be broadly classified into Hodgkin's disease (HD) and non-Hodgkin's lymphoma (NHL). Non-Hodgkin's lymphomas have been classified numerous times, but an international working formulation has adapted a fusion between all the different types of classification, which has been superseded by the REAL classification (Harris *et al.*, 1994). The major categories of the REAL classification are shown in *Table 7.5*.

Burkitt's lymphoma is a type of NHL that is endemic in children in Central Africa, although it also occurs in Europe, USA and Japan. It has a viral etiology involving the EBV. It is a very aggressive disease, the prognosis depending upon the bulk of the disease at diagnosis.

Hodgkin's disease occurs most frequently in young males, and cytology shows the presence of Reed–Sternberg cells and Hodgkin's cells. The Reed–Sternberg cells are thought to be the malignant cells, constituting up to 5% of the tumor.

7.4   Chromosome abnormalities observed in non-Hodgkin's lymphomas

The majority of NHL patients show cytogenetic changes, with highly complex karyotypes. This obviously makes accurate interpretation of the prognostic significance of the findings difficult, as the important primary change cannot be assessed with any accuracy. However, some of the changes observed have been found to be associated with certain disease types. The primary changes often involve one of the Ig loci at 14q32, 2p12 and 22q11, and a number of other chromosomal sites, resulting in juxtaposition of genes, and consequent deregulation of the Ig gene involved, as has been explained in the previous sections on acute disease. Numerical changes are usually considered to be secondary changes.

Burkitt's lymphoma

The translocations t(8;14)(q24;q32), t(8;22)(q24;q11) and t(2;8)(p11;q24) are associated with Burkitt's lymphoma, and also with B cell ALL (*Section 6.5*). The region 8q24 is common to all the translocations and this is the location of the *MYC* oncogene. The translocations also all activate *MYC* transcription.

Non-Hogkins lymphomas – follicle center involvement

Translocation t(14;18)(q32;q21) is the most common chromosome abnormality in NHL. Variant translocations t(2;18)(p12;q21) and t(18;22)(q21;q11) have also been observed. The *BCL2* gene maps to the 18q21 breakpoint, and as a result of the translocations it is brought under the influence of the *IgH* gene enhancer at 14q32, *Igκ* at 2p12 and *Igλ* at 22q11. This leads to overexpression of *BCL2*. The translocations occur specifically in B cell lymphomas of follicle center cell origin, particularly small cell type. If observed alone, they are associated with a benign clinical course, but with other changes disease progression tends to be more rapid, and in fact they are normally observed as part of a complex karyotype.

Other changes in NHL

■ Trisomy 2 is observed in follicular lymphoma, carries a poor prognosis, and is often seen with other trisomies.

■ Trisomy 18 occurs in 10–15% of NHL and is usually accompanied by additional changes. It is observed most frequently in low-grade follicular lymphoma.

Table 7.5 REAL Classification summary

B cell neoplasms including:

Precursor B lymphoblastic leukemia/lymphoma and peripheral B cell neoplasms

B cell chronic lymphocytic leukemia (CLL) and prolymphocytic leukemia (PLL)

Small cell lymphocytic lymphoma

Lymphoplasmacytoid lymphoma

Immunocytoma

Mantle cell lymphoma

Follicle center lymphoma, follicular

    Provisional cytological grades:

    (a) small cell

    (b) mixed small and large cell

    (c) large cell

    (Provisional subtype: diffuse predominantly small cell type)

Marginal zone B-cell lymphoma, extranodal MALT-type (+/– monocytoid B cells)

    (Provisional subtype: Nodal (+/– monocytoid B cells))

    Provisional entity: Splenic marginal zone lymphoma (+/– villous lymphocytes)

Hairy cell leukemia (HCL)

Plasmacytoma and plasma cell myeloma

Diffuse large B cell lymphoma (DLBCL)*

    (Subtype: Primary mediastinal (thymic) B cell lymphoma)

Burkitt's lymphoma (BL)

    Provisional entity: High grade B cell lymphoma, Burkitt-like*

T cell and putative NK-cell neoplasms including:

Precursor T lymphoblastic lymphoma/leukemia and peripheral T cell and NK neoplasms

T cell chronic lymphocytic leukemia and prolymphocytic leukemia

Large granular lymphocytic leukemia (LGL), T cell type and NK type

Mycosis fungoides and Sézary's syndrome

Peripheral T cell lymphomas, unspecified*

    (Provisional cytological categories: medium-sized cell, mixed medium and large cell, large cell, lymphoepithelioid cell)

    (Provisional subtype: Hepatosplenic γδ T cell lymphoma)

    (Provisional subtype: Subcutaneous panniculitic T cell lymphoma)

Angioimmunoblastic T cell lymphoma (AILD)

Angiocentric lymphoma

Intestinal T cell lymphoma (+/– enteropathy associated)

Adult T cell lymphoma/leukemia (ATL/L)

Anaplastic large cell lymphoma (ALCL), CD30+, T and null cell types

    Provisional entity: Anaplastic large cell lymphoma, Hodgkin's-like

Hodgkin's disease (HD)

(a) Lymphocyte predominance

(b) Nodular sclerosis

(c) Mixed cellularity

(d) Lymphocyte depletion

    (Provisional entity: Lymphocyte-rich classical HD)

*These categories are thought likely to include more than one disease entity

This table was produced with thanks to Christine Harrison of the Royal Free Hospital, London.

■ Trisomy X is a common change in NHL and is seen in 21% of follicular lymphomas.

Mantle cell lymphoma

Translocation t(11;14)(q13;q32) leads to a rearrangement of the *BCL1* gene located at 11q13 by recombination within the *IgH* locus at 14q32 leading to an overexpression of cyclin D1. This translocation is most often observed in mantle cell lymphoma but has also been found in PLL, multiple myeloma and SLVL.

T cell lymphomas

Abnormalities of 14q11

■ The region 14q11 involves the α, and δ T cell receptor loci (see *Section 7.2* in CLL). They are observed in lymphomas of T cell type.

■ Trisomy 3 is nearly always associated with a T cell phenotype and diffuse histology involving both large and small cells.

■ Abnormalities in the chromosome region 6p21–24 show a strong association with T cell lymphomas.

■ Chromosome 1 abnormalities are observed in 25% of NHL cases and are often involved in clonal evolution. The p and q arms are involved, in the form of translocations, deletions, duplications and isochromosome for 1q, the most frequent breakpoints are p22 and p36. Abnormalities of 1p are often associated with T cell lymphomas.

Anaplastic large cell lymphoma

Translocation t(2;5)(p23;q35) is observed in anaplastic large cell lymphoma, which is also known as Ki-1 positive lymphoma because of the expression of CD30 (Ki-1) antigen. The translocation fuses the nuclear protein gene, nucleophosmin (*NPM*) at 5q35 with the anaplastic lymphoma kinase gene (*ALK*) at 2p23. It is often observed as part of a complex karyotype, but is thought to carry a favorable prognosis.

Diffuse large B cell lymphoma

■ Abnormalities of 1q, particularly at q21, are associated with diffuse large B cell lymphoma (DLBCL). They are thought to carry a poor prognosis.

■ Deletions of 6q have been observed commonly in NHL in approximately 30% of cases and more specifically of DLBCL. These often occur in association with t(14;18)(q32;q21). The breakpoints are highly variable, although the regions 6q25–27 (*RMD1*) and 6q21–23 (*RMD2*) are most commonly involved.

■ Trisomy 7 is most often observed in DLBCL, is associated with a poor prognosis, and is observed in 5–15% NHL.

Small cell lymphocytic lymphoma

Breakpoints 11q23–25, particularly as deletions, are most often associated with small cell lymphocytic lymphomas. Trisomy 12 also occurs in small cell lymphocytic lymphoma, the lymphomatous counterpart of B cell CLL.

Other changes

- Material of unknown origin is often present on 14q32 and, including the abnormalities already described, this group of 14q abnormalities is observed in 50% of NHL cases.

- Abnormalities involving breakpoints 17q21–25 predominantly occur in follicle center cell lymphoma. The *BCL3* gene at 17q22 is often rearranged and is associated with transformation to a more aggressive stage of disease. Abnormalities of 17p involve deletions of 17p11 and the p53 gene.

7.5   Chromosome changes in Hodgkin's disease

The yield of metaphase spreads is usually low in these cases, and many normal cells proliferate, hence no real patterns of clinical and karyotypic association have emerged to date. When karyotypes are abnormal, they are usually complex, and ploidy changes are frequent. Changes observed in NHL have also been observed in HD which include trisomies for chromosomes 3 and 7 and deletions of 1p, 6q and 7q. Translocations t(14;18)(q32;q21) and t(2;5)(p23;q35) have also been observed in HD using molecular studies. Deletion of 4q at band q25–27 is the only abnormality to show some specificity to HD.

8.   Solid tumors

From the preceding sections, it can be seen that many chromosome aberrations have been identified and associated with varying disease types, particularly in the leukemias. Although improved techniques for the study of solid tumors have been available since the mid-1980s, the actual proportion of tumors studied cytogenetically represents only a minority of all the neoplasms studied to date. Approximately 29% of the database of recorded cytogenetic studies (Mitelman, 1998) is made up of solid tumor cases, and a number of consistent chromosomal changes have been observed in a variety of solid tumors, and some of these are shown below with some of their molecular consequences.

In general, simple chromosomal changes represent benign and low-grade malignant tumors, while complex changes are observed in aggressive and high-grade malignant tumors. There are exceptions – multiple chromosomal changes are observed in lipoma, uterine leiomyoma, and pleomorphic adenoma of the salivary gland, whereas the highly aggressive Ewing's sarcoma and synovial sarcoma are associated with sole balanced reciprocal translocations. Carey and Mertens (2000) provide us with a review on chromosome abnormalities in solid tumors.

8.1   Chromosome changes in solid tumors

- Alveolar soft part sarcoma – 17q25 aberrations.
- Atypical lipomatous tumor – ring or giant markers, showing molecular amplification of 12q.
- Clear cell sarcoma – t(12;22)(q13;q12) showing *EWS/ATF1* fusion.

- Dermatofibrosarcoma protuberans – r(17;22)(q22;q13) and t(17;22)(q22;q13), showing molecular *COL1A1/PDGFB* fusion.
- Desmoplastic small cell, round cell tumor – t(11;22)(p13;q12), showing *EWS/WT1* fusion.
- Ewing tumors – t(11;22)(q24;q12) with *EWS/FL11* fusion, t(7;22)(p22;q12) with *EWS/ETV1* fusion, t(2;22)(q33;q12) with *EWS/FEV* fusion, t(17;22)(q12;q12) with *EWS/EIAF* fusion, and t(21;22)(q22;q12) with *EWS/ERG* fusion.
- Fibromatosis/desmoid – 5q abnormalities, +8, and +20.
- Fibrosarcoma, juvenile – +8 with *ETV6/NTRK3* fusion, +11, +17, +20 and t(12;15)(p13;q25–26).
- Giant cell tumor of bone – telomeric associations of acrocentric chromosomes.
- Hibernoma – 11q13 abnormalities, showing gene losses.
- Lipoblastoma – 8q11–13 abnormalities.
- Lipoma – 12q13–15 abnormalities with *HMGIC* rearrangements, and t(3;12)(q27–28;q14–15).
- Myxoid chondrosarcoma – t(9;22)(q22;q12) showing *EWS/CHN* fusion.
- Myxoid liposarcoma – t(12;16)(q13;p11) showing *FUS/CHOP* fusion, and t(12;22)(q13;q12) showing *EWS/CHOP* fusion.
- Neurileoma – monosomy 22 and del(22q), with loss of *SCH*.
- Osteocartilaginous extosis – del(8)(q24), with loss of *ETX1*.
- Papillary renal cell carcinoma – t(X;1)(p11;q21), showing *TFE3/PRCC* fusion.
- Papillary thyroid carcinoma – inv(10)(q11q21), showing *RET/PTC* fusion.
- Parosteal osteosarcoma – ring chromosomes, showing amplification of 12q.
- Pleomorphic adenoma of the salivary gland – t(3;8)(p21;q12), showing *CTNNB1/PLAG1* fusion.
- Rhabdomyosarcoma, alveolar – t(2;13)(q35;q14), showing *FKHR/PAX3* fusion, and t(1;13)(p36;q14), showing *FKHR/PAX7* fusion.
- Synovial sarcoma – t(X;18)(p11;q11), showing *SYT/SSX1* fusion or *SYT/SSX2* fusion.
- Uterine leiomyoma – t(12;14)(q14–15;q23–24), showing *HMGIC* and *RAD51B* changes.

9.   Summary

It can be seen from this particular chapter that the study of chromosomes in malignancy is an expanding field with many areas we have yet to explore. Although we could fill many volumes with the different changes that have been observed in disease, we hope this chapter has provided the reader with a basic knowledge of the more established abnormalities and disease associations.

With all the modern technologies at our disposal (see *Chapter 8*), there may prove to be further success stories, such as the tailored treatment for the t(15;17) based on the gene disruption involved in the translocation. As it has been said, 'You ain't seen nothing yet!'

Acknowledgments

This chapter is dedicated to all those who work with chromosomes in malignancy as they can be so frustrating to analyze! I would like to give particular thanks to Tony Potter and Ann Whatmore of the Centre for Human Genetics at Sheffield, Barbara Gibbons of Great Ormond Street Hospital, London, and Christine Harrison of the Royal Free Hospital, London. These steadfast souls have always provided me with smiles, laughter and inspiration along the way during my work with malignant tissues.

References

Bain, B.J. (1990) *Leukaemia Diagnosis: a Guide to the FAB Classification.* Gower Medical, London.

Bennett, J.M., Catovsky, D., Daniel, M.T., Flandrin, G., Galton, D.A.G., Gralnick, H.R. and Sultan, C. (1982) *Br. J. Haematol.* **51**: 189.

Bennett, J.M., Catovsky, D., Daniel, M.T., Flandrin, G., Galton, D.A.G., Gralnick, H.R. and Sultan, C. (1985) *Ann. Intern. Med.* **103**: 620.

Carey, T. and Mertens, F. (2000) *Clinical Laboratory Medicine*, 2nd edn (ed. K. McClatehey). Lippincott Williams and Wilkins, Philadelphia, PA.

First MIC Cooperative Study Group (1986) *Cancer Genet. Cytogenet.* **23**: 189.

Fourth International Workshop on Chromosomes in Leukaemia, 1982 (1984) *Cancer Genet. Cytogenet.* **33**: 254.

Harris, N.L., Jaffe. E.S., Stein H., Banks, P.M., Chan, J.K., Cleary, M.L., Delsol, G., De Wolf-Peeters, C., Falini, B. and Gatter, K.C. (1994) *Blood* **84**: 1361–1392.

Juliusson, G., Oscier, D.G., Fitchett, M., Ross, F.M., Stockdill, G., Mackie, M.J., Parker, A.C., Castoldi, G.L., Cuneo, A., Knuutila, S., Elonen, E. and Gahrton, G. (1990) *New Engl. J. Med.* **323**: 720–724.

Mitelman, F. (1998) *Catalog of Chromosome Aberrations in Cancer '98*, CD-ROM, Version 1. Wiley-Liss, New York.

Nowell, P.C. and Hungerford, D.A. (1960) *Science* **132**: 1497–1499.

Seabright, M. (1972) *Chromosoma* **36**: 204.

Secker-Walker, L.M. (1998) *Leukaemia* **12**: 776.

Second MIC Cooperative Study Group (1988) *Br J. Haematol.* **68**: 487.

Further reading

Mufti, G.J., Flandrin, G., Schaefer, H-E., Sandberg, A.A. and Kanfer, E.J. (1996) *An Atlas of Malignant Haematology, Cytology, Histology and Cytogenetics.* Martin Dunitz, London.

Chapter 8

New complementary techniques

Karen Saunders and David Jones

1. Introduction

Conventional cytogenetic studies are limited by the fact that they require dividing cells to observe the chromosomes. There is also a limit to the resolution that can be achieved using conventional chromosome analysis. As already mentioned in *Chapter 4*, the resolution of the light microscope only allows the identification of chromosome changes that are 4 Mb or larger in size. Maximum resolution is dependent on the quality of the preparations and in poor-quality preparations the resolution is greatly reduced. Small deletions or duplications may therefore be impossible to see. Problems also arise when the chromosomes become rearranged. Often the rearranged segments can be identified from the G-banding pattern. However, this is not always possible, particularly in the case of small supernumerary marker chromosomes.

Over the years complementary techniques have been developed to allow the clarification and identification of chromosome abnormalities that cannot otherwise be achieved using conventional cytogenetics. Techniques have also been developed to try and bridge the gap between nonmolecular cytogenetic and molecular DNA techniques, and allow the visualization and identification of smaller segments of DNA. Consequently, even more information about cytogenetic abnormalities can now be obtained. This chapter describes some of the most important techniques that are currently in use.

2. Fluorescent *in situ* hybridization

2.1 Introduction

Fluorescent *in situ* hybridization (FISH) is a technique that uses molecular DNA probes to detect complementary DNA sequences along a chromosome. The single-stranded probe DNA is allowed to anneal (hybridize) to the single-stranded target DNA, which is still in its natural position on the chromosome, *in situ*. The probe is coupled to a fluorochrome, a molecule which fluoresces when it is excited by a particular wavelength of light, and it is visualized using a fluorescence microscope. Several different probes can be hybridized, observed and compared simultaneously by using different colors and combinations of fluorochromes for each probe. FISH allows the simultaneous assessment of molecular and cytogenetic information on metaphase chromosome preparations. It may also be used on interphase cells to provide information when the cells are not dividing and chromosome preparations cannot be obtained. It has, therefore, become an invaluable tool in the diagnostic and research laboratory.

2.2   Applications of FISH

FISH can be used to analyze numerical and structural aberrations. It provides a very rapid and specific test that is able to simultaneously detect different nucleic acid targets within the same preparation. The most common uses include the characterization of rearrangements and marker chromosomes, the detection of numerical abnormalities, and the detection of microdeletions, duplications and structural rearrangements that are beyond the resolution of the light microscope.

2.3   Important steps in FISH

FISH involves several steps before the probes may be visualized. However, the process is continually becoming more efficient and results can now be achieved in just a few hours.

Slide preparation

Slides are usually prepared in a similar way to that already described for routine cytogenetic analysis (see *Chapter 4*). However, FISH may also be carried out on slides of formalin-fixed tissue, blood or bone marrow smears, and uncultured, fixed cells. The slides may be pre-treated, if necessary, in a protease solution to make the cells more permeable to the probe. This step can greatly increase the efficiency of hybridization for samples that are uncultured and/or particularly cytoplasmic, such as uncultured amniocytes. Hybridization also appears to be more efficient on slides that have been slightly aged, either on the bench for a couple of days or by treating with a $2 \times$ salt and sodium citrate solution (SSC) at $37°C$ for about 30 min. The slides are then dehydrated through an ethanol series, which helps make the cells more permeable.

Denaturation of probe and target DNA

Before the probe and target DNA can be hybridized together, they have to be rendered single-stranded, which is achieved by heating them. The probe can be added to the slide with the target DNA and they can be denatured together on a hotplate. Alternatively the probe and target DNA can be denatured separately using solutions such as formamide and salt solutions which lower the temperature at which the DNA denatures and, therefore, helps to maintain the structure of the chromosomes more effectively. In some cases the probes are already single-stranded, and as such do not require denaturing.

Hybridization

Once the probe and target DNA have been denatured and added together, the probe DNA must then be allowed to anneal to the target DNA. This is usually carried out at $37°C$, which is the optimum temperature for hybridization. At higher temperatures, the rise in energy increases the rate of movement of the probe and therefore reduces the time it takes to find the target DNA. The bonds formed between the probe and target DNA (the duplex molecules) are weaker, causing them to dissociate, which ultimately results in a decrease in hybridization. At lower temperatures there is less energy for the movement of the probe and hybridization is slower.

Post-hybridization wash

During hybridization, the probe will bind to the specific target DNA but some copies of the probe DNA will also bind to less specific stretches of DNA in the sample. Therefore, if the sample was to be observed immediately after hybridization there would be a lot of nonspecific hybridization signals, otherwise known as background, and the specific probe signal would be impossible to distinguish. In order to remove the background hybridization the sample has to be washed in a stringent solution, after hybridization, which breaks the bonds of the nonspecifically bound probe and washes it away, leaving the specifically bound probe intact.

The stringency wash works by taking advantage of the fact that the more hydrogen bonds that form between the DNA strands, the more energy is required to break them up. Therefore, the closer the match to the target DNA, the more hydrogen bonds are formed between the probe and target DNA and, hence, the stronger the hybridization. The stringency of the post-hybridization wash is therefore designed to dissociate imperfect matches, whilst leaving the perfect and, therefore, most strongly hybridized matches intact. Stringency is affected by temperature and salt concentration. An increase in temperature will increase the dissociation of the hybridization due to the increase in energy. Increase in salt concentration will decrease the stringency since it increases the amount of energy required to break the hydrogen bonds. Other substances may also be used, such as formamide, which increases the stringency by lowering the dissociation temperature of the DNA.

Detection

There are two types of detection systems, direct and indirect, as described below.

(a) Indirect detection The probe used is coupled to a reporter molecule such as biotin or digoxygenin. In order to detect the probe, a fluorochrome is conjugated to a molecule, which binds to the reporter molecule. An additional step is therefore required after hybridization and stringency washing. This method has the advantage that several layers of fluorochrome can be added to amplify the fluorescent signal, if required, making it very sensitive for probe detection.

The biotin reporter molecule used in FISH is a B vitamin, which has a strong binding affinity with the protein avidin. Biotin can, therefore, be detected using a fluorochrome labeled with avidin. If the signal appears weak it may be amplified by adding an antiavidin antibody which is biotinylated (conjugated to a biotin molecule). This allows another layer of fluorochrome labeled avidin to be added. It is an open-ended system, and the signal can therefore be amplified as many times as desired. However, this will also increase the detection of any background signals and so a balance has to be found between achieving a bright probe signal without detecting too much background fluorescence.

The reporter molecule digoxygenin is detected using antidigoxygenin antibodies conjugated to a fluorochrome. The signal can be amplified using yet another antibody to the one already added. For example if the first antibody used is antidigoxygenin that has been raised in a mouse, the second layer will have to be an anti-mouse antibody produced in a different animal such as a rabbit. The signal can be amplified as many times as there are antibodies available, but again the balance has to be found between achieving a bright signal without detecting excessive background. In practice, for most hybridizations, one layer of fluorochrome is sufficient to produce a good signal.

(b) Direct detection No extra detection step is required. The probe is directly coupled to a fluorochrome and can therefore be observed immediately after hybridization and stringency washing. Many commercial probes are now available that are directly labeled for ease of use.

Counterstaining

The chromosomes and/or the nuclei on the slide have to be counterstained before examination so that the relative position of the probe can be visualized. The most popular counterstains are DAPI, which is blue, and propidium iodide, which is red. The choice depends on the color of probes being used. Obviously, if a red fluorescently labeled probe is to be visualized then propidium iodide would be an inappropriate counterstain.

Microscopy and digital image production

In order to visualize the probes, an epifluorescence microscope fitted with a 200 W mercury light bulb, which provides ultraviolet light, and a set of filters are usually used for routine FISH analysis. Different fluorochromes have different excitation wavelengths of light and different filters are therefore required to produce them. The light emitted by the fluorochrome will be of a different wavelength to that of the excitation light and a second set of filters are required to block the excitation light and allow only the emission wavelength of the fluorochrome to reach the observer. Examples are shown in *Table 8.1*.

Filter sets may also be used that allow the simultaneous detection of two or more fluorochromes, for example dual and triple band pass filters. These can be useful when different probes need to be observed in relation

Table 8.1 Excitation levels of fluorochromes

Fluorochrome	Excitation (nm)	Emission (nm)
DAPI (blue)	360	490–500
Propidium iodide (red)	340	600–610
FITC (green)	490	525
Rhodamine (red)	596	615

to one another for the detection of rearrangements. However, these filters have the disadvantage that the signal brightness for each probe is diminished and they are best used in combination with the single band pass filters as well.

Fluorochromes fade over time and upon exposure to light. It is, therefore, an advantage to be able to produce images immediately so that permanent records of the tests carried out may be stored. By using a camera linked to a computer system with special software, digitized images can be produced. The image can then also be manipulated to enhance the probe signal and reduce background. This image can then be stored and hard copies produced if required.

Analysis

The results are analyzed and interpreted by eye. Analysis usually involves the observation of fluorescent signals either on metaphase chromosomes or in interphase nuclei. The interpretation depends on the probe(s) used and whether signals are present, absent or repositioned. This is usually fairly straightforward in metaphase cells. However, hybridization may not always have worked perfectly in all cells and so sometimes many cells have to be examined before a conclusion can be reached. This is particularly the case for interphase analysis, which is complicated by the fact that the nucleus does not lay completely flat on the slide. It can therefore produce a slightly three-dimensional effect and the microscope lens often has to be focused up and down in order to observe all the probe signals. In addition, probe signals which are spatially separate, may occasionally appear very close to one another, or, even on top of each other, and consequently, only one signal may be observed when actually two are present. Conversely, splitting of signals sometimes occurs and two signals may be observed when actually only one locus is present. Because of this, each laboratory must establish the normal range of signals observed in normal cells, for each probe used, in order to recognize an abnormal hybridization pattern. Many cells, therefore, have to be analyzed for interphase studies before a conclusion can be reached, particularly for aneuploid diagnosis, and it can be difficult or impossible to detect mosaic aneuploidy in interphase cells.

Types of probe

There are four main types of probe that are commonly used for clinical FISH studies, which are described below.

(a) Alpha satellite probes These probes are made up of the repetitive DNA sequences found in the alpha satellite centromeric region of the chromosome and typically consist of about 10 repeated base pairs. They are chromosome specific, with two exceptions: chromosomes 13 and 21 share the same sequences, and so do chromosomes 14 and 22. Alpha satellite probes produce very bright signals and are particularly useful for chromosome enumeration in interphase nuclei and for identifying marker chromosomes in metaphase preparations. Because the sequences are often quite similar between the chromosomes, they can sometimes produce problems with cross-hybridization and therefore the stringency of the post-hybridization wash has to be very well controlled. Centromere probes are also available which include sequences common in every chromosome alpha satellite region and will therefore fluoresce the centromeres of all the chromosomes present (see *Figure 8.1* in the color section between pages 18 and 19).

(b) Telomeric probes These probes are made up of the repetitive DNA sequences found in the subtelomeric region of the chromosome. They are chromosome-specific and specific for the p or q arm of the chromosome. Telomeric probes are particularly useful for identification of cryptic terminal chromosome deletions and rearrangements. Telomere probes are also available which include sequences common to all the chromosomes telomeric regions, causing the telomeres of all the arms of all the chromosomes to fluoresce.

(c) Unique sequence probes These probes range from very small single copy probes such as PCR products to cosmids (cloning vectors which can accommodate foreign DNA fragments between 30 and 44 kb in size) and yacs (yeast artificial chromosomes). The smaller the probe the fainter the signal observed will be. The optimum probe size is about 5 kb; probes smaller than 1.5 kb, may fail to show a signal. Cosmid probes are the most commonly used unique sequence probes. Unique sequence probes are locus-specific and can detect a specific chromosome region or gene. They are especially useful in identifying some microdeletion syndromes such as DiGeorge syndrome at chromosome region 22q11.2, Williams syndrome (Elastin gene) at chromosome region 7q11.23 and the Prader–Willi/ Angelman syndrome at chromosome region 15q11–q13, since these abnormalities can be very difficult or impossible to observe on conventional cytogenetic preparations. Micro-deletion region probes are usually used in conjunction with a control probe of a different color to help distinguish true deletion of the probe from failure of the probe to hybridize. Unique sequence probes are also increasingly being used to identify cryptic rearrangements in samples from leukemia patients. The most commonly used probes in this instance are the *BCR* and *ABL* locus probes, which are used to detect the *BCR/ABL* fusion gene observed in chronic myeloid leukemia. By using a *ABL* locus probe, located on the long arm of chromosome 9 at q34, and an *BCR* locus probe, located on the long arm of chromosome 22 at q11.2, labeled with different color fluorochromes, the fusion gene may be observed in metaphase or interphase cells (see *Figure 8.2* in the colour section between pages 18 and 19). Finally, the use of a locus-specific probe for the Down syndrome critical region on chromosome 21 is increasingly being used to rapidly detect trisomy for this region, and hence Down syndrome, on uncultured amniocytes.

(d) Whole chromosome paints These consist of a mixture of unique sequence probes from a single chromosome, which are derived from flow-sorted chromosome-specific libraries. These probes, as the name suggests, 'paint' or fluoresce the entire length of a metaphase chromosome. They are used to detect cryptic translocations and to identify the origin of unidentified material on rearranged chromosomes or marker chromosomes (see *Figure 8.3* in the color section between pages 18 and 19).

2.4 Reporting of FISH results

As for routine cytogenetic studies, there is also an international system to describe FISH findings in a concise and accurate way, the guidelines of which are set out in the 1995 edition of the International System of Chromosome Nomenclature (ISCN). It uses symbols and abbreviations to convey information on the type of probe used, its location, and what the observed results were. It can also be used in conjunction with standard cytogenetic nomenclature to combine information that has been detected using nonmolecular cytogenetics with information that has subsequently been obtained using FISH. The two sets of nomenclature are separated using a full stop (.) with the nonmolecular cytogenetic result being placed first.

Some examples of the more commonly used symbols and abbreviations are shown below; for a fully comprehensive list refer to the 1995 ISCN.

−	the probe signal is absent from a specific chromosome
+	the probe signal is present on a specific chromosome
++	duplication of the signal on a specific chromosome
×	precedes the number of signals seen
.	separates cytogenetic observations from results of *in situ* hybridization
;	separates probes on different derivative chromosomes
ish	*in situ* hybridization when used on metaphase chromosomes
mv	the probe signal has moved from its original location due to a rearrangement
st	stationary signal with the probe remaining in its original position
nuc ish	interphase *in situ* hybridization
wcp	whole chromosome paint

A few basic examples of the use of the standard nomenclature for FISH are shown below, with explanations of the results shown.

(a) Metaphase in situ hybridization (ish)

In situ hybridization on metaphase chromosome preparations is described by abnormality type, for example, duplication (dup), deletion (del), derivative chromosome (der), and the number of the chromosome being referred to (as used in standard cytogenetic ISCN), followed by a description of the probe used and the signal observed, that is whether the signal is present, absent or duplicated, in brackets:

$$46,XY.ish\ del(22)(q11.2q11.2)(D22S75-)$$

This demonstrates that conventional cytogenetics was carried out, and the karyotype was found to be 46,XY, an apparently normal male. FISH was then carried out on a metaphase chromosome preparation using the probe D22S75 which is located on the long arm of chromosome 22 within band 11.2. The result identified a deletion (absence) of the probe signal on one of the chromosome 22s.

If more than one probe is being investigated, then they are listed in order from the short arm of the chromosome to the long arm of the chromosome and separated using a comma:

$$46,XY.ish\ del(15)(q11.2q11.2)(SNRPN-,D15S10-)$$

This symbolizes that FISH identified that two different probes, SNRPN and D15S10, which map within band 11.2 on the long arm of chromosome 15, were deleted.

The number of signals observed on normal chromosome homologs are represented by a multiplication sign followed by the number of signals seen. The chromosome location is not included in parentheses:

$$46,XY.ish\ 22q11.2(D22S75\times2)$$

The probe D22S75, located on the long arm of chromosome 22 at band q 11.2 (22q11.2), showed a signal on both chromosome 22s. This allows a distinction between the number of signals observed on two normal homologous chromosomes and an additional probe signal due to a duplication of the chromosome locus, which would be written as shown below:

$$46,XY.ish\ dup(22)(q11.2q11.2)(D22S75++).$$

Whole chromosome paints are referred to in the nomenclature as wcp followed by the specific chromosome that the paint detects, for example wcp5 is a whole chromosome paint for chromosome 5.

$$46,X,+\ r.ish\ r(X)(wcpX+)$$

In the example above, conventional cytogenetic analysis detected 46 chromosomes, but only one sex chromosome, the X chromosome, and an unidentified ring chromosome. FISH, using a whole X chromosome paint, confirmed that the ring chromosome was derived from the X chromosome (wcpX+).

The information on probes, located on different derivative chromosomes involved in a rearrangement, is separated using a semicolon:

$$ish\ t(9;22)(ABL-;BCR+,ABL+).$$

Using the *ABL* probe, which is normally located on chromosome 9, and the *BCR* probe, which is normally located on chromosome 22, FISH identified a translocation between chromosome 9 and chromosome 22. An *ABL* probe signal was absent from chromosome 9 and was observed instead on chromosome 22, distal to the *BCR* probe signal.

(b) Interphase/nuclear in situ hybridization (nuc ish)

The information obtained from interphase FISH includes the number of signals observed from a probe and the relative position of different probe signals. The number of signals is indicated using the abbreviation 'nuc ish' followed by the chromosome band location of the probe locus, and the locus designation, a multiplication sign and the number of signals seen in parentheses. Commas are used to separate the information on different probe loci. Different band designations of probe loci are listed in order, starting from the end of the p arm to the end of the q arm, and loci on different chromosomes are ordered starting with the sex chromosomes followed by the autosomes 1–22:

$$nuc\ ish\ Xp22.3(KAL\times2),21q22(D21S65\times2,D21S64\times1),$$
$$21q22.3\text{-}qter(D21S1219\times2).$$

Interphase FISH showed that there were two copies of the *KAL* locus, which is situated on the short arm of the X chromosome at band 22.3. In addition, two copies of the *D21S65* locus, and one copy of the *D21S64* locus situated on the long arm of chromosome 22 at band q22, and also two copies of the *D21S1219* locus situated on the long arm of chromosome 22 at band q22.3 are present.

The signals from loci on two separate chromosomes will be randomly spatially situated within each nucleus examined. However, if they have become juxtaposed next to each other on the same chromosome because of a translocation, then the signals from the loci will also appear to be juxtaposed within the nucleus. This is indicated using the abbreviation 'con'. For example, this is the case with the *BCR/ABL* rearrangement in chronic myeloid leukemia and, by using different colored fluorescent probes for these loci, in the same assay, the rearrangement can be observed in interphase nuclei. The results are recorded as follows. A normal result, with no *BCR/ABL* rearrangement is:

nuc ish 9q34(ABL×2),22q11(BCR×2).

Two separate signals were observed from the *ABL* locus probe on 9q34 and two separate signals were observed from the *BCR* locus probe on 22q11.

An abnormal result, that is translocation of the *ABL* locus from one chromosome 9 to the *BCR* locus on one chromosome 22 is shown below:

nuc ish 9q34(ABL×2),22q11(BCR×2)(ABL con BCR×1).

Two signals were observed from the *ABL* locus probe on 9q34 and two signals were observed from the *BCR* locus probe on 22q11. However, one of the *ABL* probe signals was juxtaposed to one of the *BCR* probe signals.

These are just a few examples of the use of the standard nomenclature for reporting FISH results. This nomenclature is often used just as a quick reference for the experienced cytogeneticist and once the result has been recorded according to the ISCN, a written report is also included. This explains the findings, how they relate to the referral reason, what the implications of the result are and whether any other family or follow-up studies are required.

2.5 Protocols

Different probes require slightly different conditions and, with commercial probes, detailed protocols are always provided by the manufacturer. All methods will basically include the steps already described above and it is up to each laboratory to adapt their methods to the probes that they use and the work that they routinely carry out. Some examples of protocols that can act as a guide are shown at the end of this chapter.

3.   Twenty-four color painting

As the title suggests, techniques now exist that enable the simultaneous painting of metaphase chromosomes, in 24 different colors using FISH and specific chromosome paints. Each chromosome from 1 to 22, and the sex chromosomes, X and Y, can be distinguished from each other on the basis of their color. The different colors are achieved using a combination of five or more fluorochromes. There are two different ways in which the results can be observed, and both require computer systems equipped with appropriate software.

The first method acquires an image using an epifluorescence microscope and a set of six filters mounted in a motorized wheel: five filters for the fluorochromes and one for the DAPI counterstain. Six images are automatically captured using a charged coupled device (CCD) camera, which takes about 20 s. The true colors of the chromosomes are hard to distinguish by eye. However, the computer program is able to distinguish them by their individual color wavelength and can produce a karyotype on this basis. It can then digitally false color each chromosome pair in order to increase the contrast between each one so that they can be easily recognized by eye. In addition, the computer is able to produce a karyotype on the basis of each chromosome's color variation.

The second method, known as spectral karyotyping, uses an inferometer to send a signal to a CCD camera that takes a black and white image. Hundreds of images are taken and combined to produce a spectral image, which can take several minutes. The computer then processes the resulting image and the chromosomes are distinguished from one another based on the differences in their spectra. As above, the computer program can then digitally false color each chromosome and produce a karyotype (see *Figure 8.4* in the color section between pages 18 and 19) (Schroek *et al.*, 1996).

Both methods produce very similar results and can recognize and karyotype the chromosomes relatively accurately. Twenty-four-color karyotyping is particularly useful in analyzing very abnormal cells such as tumor cells. These cells often contain a lot of derivative chromosomes and marker chromosomes, which are impossible to identify using routine cytogenetic analysis but can be partly identified using 24-color painting simply by their color.

Although these methods can be extremely useful they also have limitations. Twenty-four-color painting cannot identify all chromosome rearrangements, for example, inversions and small deletions. Small rearrangements may also not be detected due to a slight overlap of the fluorescent signal from the adjacent part of the chromosome across the rearranged segment and/or a lack of sensitivity at the telomeric regions. Finally, each assay is relatively expensive and time consuming to perform so, as a test, it is beyond the reach of most diagnostic laboratories.

4.   Color banding

A color banding pattern may be achieved on human chromosomes using subregional paints. These are obtained using whole chromosome paints derived from gibbon chromosomes. When these paints are hybridized to human chromosome preparations, they hybridize to conserved sequences on different regions of the human chromosomes, which results in each human chromosome showing a unique hybridization pattern. By combining the gibbon paints with different fluorophores, a relatively simple multicolor bar code along the chromosome can be produced. Other primate species are also currently under investigation, and could be used to further increase the color band resolution by mixing probes derived from them with the probes from the gibbon species. The color bands produced by this method allow quick chromosome identification, which can be easily automated using specialized computer programs. It may also be used to complement the G-banding pattern in the cases of chromosome rearrangements, duplications and deletions to aid the further identification of the chromosome segments involved.

5.   Primed *in situ* hybridization

Primed *in situ* hybridization (PRINS) is an alternative technique to standard FISH. In this method, short DNA fragments (primers) are hybridized to the denatured test DNA. Nucleotides that have been labeled with biotin or digoxygenin are then added and, in the presence of DNA polymerase, extend the primer DNA strand. The labeled nucleotides are then detected using a fluorescently labeled antibody. Primer extension can be carried out in about 30 min; however, the sensitivity of the technique can be improved by using the polymerase chain reaction (PCR) to increase the signal intensity. This is referred to as cycling PRINS and can produce very sensitive and specific results. By using appropriate primers, may be identified differences in primer sequence or small deletions, which are observed as the presence or absence of signals. Because of its specificity it is less prone to background fluorescence than standard FISH.

6.   Fiber FISH

Fiber FISH is a technique where *in situ* hybridization is carried out on extended DNA fibers, obtained from interphase nuclei, to produce a high-resolution physical DNA map. Duplex DNA is stretched across a microscope slide, then denatured and hybridized with DNA probes labeled either with biotin or digoxygenin and detected using fluorescent avidin or antibodies (Parra *et al.*, 1993), in much the same way as in standard FISH procedures. The results are observed using a fluorescence microscope. This technique therefore produces a high-resolution DNA map that can be directly visualized. The probe loci can be hybridized to the strand in order to map the number of loci along the strand and, when more than one probe is used, the relative position of them determined.

7.   Comparative genomic hybridization

Comparative genomic hybridization (CGH) is a technique which can be used on all types of malignant tissue whereby the whole genome is tested to determine changes in DNA sequence copy number, that is amplifications and deletions. The procedure involves labeling tumor test genomic DNA and normal genomic reference DNA with different fluorochromes. The two genomic DNA sequences are then simultaneously hybridized to normal metaphase chromosomes in the presence of Cot-1-DNA in order to block the repetitive sequences.

The ratio of fluorochrome intensity along the length of a chromosome is proportional to the ratio of tumor and control DNA hybridized at each point, due to the independent hybridization kinetics of both the test and reference DNA. A higher ratio of tumor DNA at a particular point indicates an amplification of the test DNA sequence. Conversely, a higher ratio of control DNA indicates a deletion of the test DNA sequence.

The advantages of CGH are outlined below.

■ It does not require fresh tumor tissue, since DNA may be extracted from frozen or fixed material, eliminating the requirement for cell culture.

■ It precludes the requirement for cytogenetic analysis of dividing cells, which may be limited in certain tumor tissues due to the technical difficulties in obtaining metaphases and interpretation of complex karyotypes. Also, cultured cells may not be a true representation of the neoplastic tissue under investigation.

■ Unlike FISH, where the detection of amplifications and deletions is limited to the sites of particular sequence-specific probes, that is targeting a tiny portion of the genome, CGH can map imbalances throughout the entire genome in a single hybridization event, and therefore identify locations of potential interest.

■ Amplifications can be mapped to a resolution of at least 10 Mb and high copy number amplifications can be mapped to 2 Mb.

There are, however, a number of limitations with this technique that should be understood:

■ the sensitivity of the method for detecting low copy number changes and small imbalances is limited;

■ the minimal deletion size detected with CGH is between 10 and 20 Mb.

Bentz *et al.* (1998) showed that, in a series of patients with lymphoproliferative disorders, deletions of 10–20 Mb on the long arm of chromosome 11 could only be detected in patients with a clone size of greater than 80%. When patients had smaller levels of clone size, that is less than 60%, small deletions could not be detected. The level of contaminating normal cells can be important to the sensitivity of the technique, particularly in identifying smaller deletions. Therefore, both criteria must be considered for accurate interpretation of CGH results.

Many neoplasms, in particular hematological malignancies, may have a relatively small tumor load. Moreover the identification and mapping of chromosomal abnormalities have focused on increasingly smaller regions. These limitations have restricted the applications of CGH.

Another important consideration is that CGH yields no information on chromosomal translocations, inversions and insertions, and cannot distinguish between diploid and tetraploid tumors. Therefore, it is usually only fully informative when used in conjunction with other molecular and nonmolecular cytogenetic techniques.

7.1   Overview of the techniques used in CGH

The methods used in CGH are time-consuming, involved and require expensive, specialized computer software for quantitative interpretation. As such, full detailed protocols are beyond the scope of this volume. However, CGH kits are commercially available, complete with full protocol lists. Outlined below are the general principles involved with this technique.

The test genomic DNA is extracted using standard laboratory techniques for DNA extraction (Alers *et al.*, 1999). The test DNA is directly labeled with fluorescently labeled dUTP by nick translation using standard laboratory techniques. The size of the test probe is then determined by running an agarose gel. The optimal size of labeled probe is 300–3000 bp for CGH experiments.

The differentially labeled test and reference genomic DNA are simultaneously hybridized to normal metaphase spreads using standard FISH techniques as outlined in *Protocol 8.4*. The slides are then subjected to post-hybridization washes and counterstained with DAPI II. The slides are then examined using a fluorescence microscope equipped with the appropriate triple-band-pass filter set in conjunction with a CGH imaging software package.

7.2   Reporting of CGH results

Guidelines for the recording of CGH findings are set out in detail in the 1995 edition of the ISCN under the heading 'Reverse *in situ* hybridization (rev ish)'.

Chromosomes or chromosome segments with enhanced (enh) or diminished (dim) fluorescence intensity ratios indicate a relative increase or decrease of the copy number, respectively, with reference to the basic euploid state. Some examples of basic rev ish nomenclature are shown below.

47,XY,+mar.rev ish enh (11p)

In a similar manner to the FISH nomenclature, the karyotype obtained from conventional, nonmolecular techniques is listed first, and separated from the CGH study results by a full stop (.). The marker chromosome has been shown by CGH to be composed mostly or wholly of material from the short arm of chromosome 11, that is 11p as shown in parentheses.

46,XX,add(7)(q36).rev ish der(7)t(7;12)(q36;q24) enh(12q24)

The above example shows that the additional material on the long arm of chromosome 7 is made up of material from the long arm of chromosome 12 at band q24.

rev ish enh (8)

The simple CGH reference above informs us that there are extra copies of chromosome 8.

rev ish dim (5q31q33)

This notation shows that there is a reduced amount of genomic material from 5q31 to 5q33.

8.   The present and future of cytogenetics

The emergence of so many complementary techniques to cytogenetics has greatly helped to identify and clarify chromosome rearrangements that would have been otherwise impossible, or very difficult, in the past. They have become of particular importance in the research environment where they enable more accurate localization of genes involved in genetic disease. These techniques are constantly developing and becoming increasingly refined, promising exciting possibilities for the future. They can provide answers to specific questions; however, as the title suggests, they are generally complementary to, and on their own cannot replace, conventional cytogenetics. In addition, many of them are not recognized as official diagnostic tests and are classified as being for research only. Despite this, they are often of great value in a diagnostic laboratory, although the results obtained from them have to be reported cautiously and the nature of the test explained in the report.

At the moment conventional cytogenetic analysis still provides the most information on the karyotype of a metaphase cell. The chromosomes are examined together and no specific questions have to be asked in order to be able to identify all ranges of chromosomal abnormality. Therefore, in the foreseeable future there can be no substitute for the trained eye of a cytogeneticist.

References

Alers, J.C., Rochat, J., Krijtenburg, P.J., van Dekken, H., Raap, A.K. and Rosenberg, C. (1999) *Genes Chromosomes Cancer* **25(3)**: 301–305.
Bentz, M. *et al.*, (1998) *Genes Chromosomes Cancer* **21**: 172–175.
ISCN (1995). *An International System for Human Cytogenetic Nomenclature.* Karger, USA.
Parra, I. *et al.*, (1993) *Nature genetics* **17**: 21.
Schroeck, E. *et al.* (1996) *Science* **26**: 273 (5274), 494.
Speicher, M.R. *et al.* (1996) *Nature Genetics* **12**: 368–75.

Protocol 8.1

Processing of uncultured blood and bone marrow preparations for interphase FISH

Equipment

Centrifuge set at 200 **g** with swinging bucket rotor

Microscope slides with frosted ends

Plastic Pasteur pipettes

Sterile centrifuge tubes, 5 ml round bottom

Solutions

Fixative: three parts Analar-grade methanol to one part Analar-grade acetic acid

Potassium chloride, 0.75 mM (KCl)

Protocol
Sample preparation

1. Add 1 ml of 0.75 mM KCl to 0.25–1 ml of blood or bone marrow in a centrifuge tube.

2. Incubate for 10 min at 37°C.

3. Centrifuge for 5 min.

4. Remove the supernatant and slowly add 5 ml fixative.

5. Centrifuge and replace the fixative two to three times.

Slide-making

1. Clean microscope slides by dipping in methanol and drying with a tissue.

2. Centrifuge the fixed cell suspension for 5 min at 200 **g**.

3. Remove the supernatant.

4. Add a drop of the suspension onto a microscope slide.

5. Allow to air dry.

6. Check the cell density under phase contrast.

Protocol 8.2

Processing of uncultured amniocytes for interphase FISH

Equipment

Acrodisc filter 0.2 µl

Centrifuge tubes 5 ml sterile

Coplin jars

Gilson pipette

Microscope slides with frosted ends

Pasteur pipettes, 1 ml

Syringe sterile, 5 ml

Waterbath

Solutions

Fixative: three parts Analar-grade methanol, to one part Analar-grade acetic acid

Potassium chloride, 60 mM (KCl)

$2 \times SSC$

Trypsin/0.53 mM EDTA 0.5%

Protocol

1. Pre-warm the trypsin/EDTA, 60 mM KCl, $2 \times SSC$ and the trypsin solution to 37°C.

2. Centrifuge 3–5 ml of amniotic fluid in a centrifuge tube for 5 min at 200 *g*.

3. Tip off supernatant and resuspend the cell pellet in 2ml of warm 1 × trypsin/ EDTA that has been filtered through an acrodisc using a 5 ml syringe to remove any bacteria that could be present.

4. Incubate for 10 min in the waterbath at 37°C.

5. Centrifuge for 5 min at 200 *g*.

6. Remove the trypsin/EDTA and resuspend the cell pellet in 2 ml of 60 mM KCl.

7. Incubate for 10 min in a waterbath at 37°C.

8. Slowly add 2 ml of fixative to the cells in hypotonic solution; do not mix.

9. Centrifuge for 5 min at 200 *g*, remove the supernatant and add 2 ml of fresh fixative.

10. Mix this fixed cell suspension with a pipette.

Slide-making

1. Clean the microscope slides by dipping in methanol and drying with a tissue.

2. Centrifuge the fixed cell suspension for 5 min at 200 *g*.

3. Remove the supernatant.

4. Using a Gilson pipette, draw up about 40 μl of cell suspension.

5. Run out the suspension onto a microscope slide.

6. Allow the slide to air dry tipped upright.

Protocol 8.3

Pre-treatment of slides for uncultured samples (or very cytoplasmic samples)

This method uses a pepsin solution to gently digest away cytoplasm and cellular proteins that may otherwise reduce the efficiency of hybridization. The slides are then treated with a magnesium chloride and formaldehyde solution to re-fix the nucleic acids.

Equipment

Coplin jars

Pasteur pipettes, 1 ml

Waterbath

Solutions

Absolute ethanol, Analar-grade

Formaldehyde solution (40%)

Hydrochloric acid (HCl) (1.0 M)

Magnesium chloride 6-hydrate

Pepsin 4150 units (mg solid)$^{-1}$

Phosphate buffered saline solution (PBS)

Sodium hydroxide (0.1 M)

Sodium saline citrate (SSC), 20 × SSC buffer

Make-up of the solutions required

70, 90 and 100% ethanol: make the dilutions up with distilled water.

2 × SSC, add 500 ml 20 × SSC to 4.5 l of distilled water. Adjust to pH 7 using concentrated HCl or sodium hydroxide (use about 5 ml to adjust for every pH point in error).

0.4 × SSC: add 10 ml of 20 × SSC to 500 ml of distilled water. Adjust to pH 7 using concentrated HCl or sodium hydroxide.

HCl solution: add 400 μl of concentrated HCl (1.0 M) to 500 ml of distilled water.

Formaldehyde solution: add 2.25 g of Mg_2Cl_2 to 12.5 ml of formaldehyde (40% solution) and make up to 500 ml using PBS.

Pepsin working solution: add 80 μl of 10% pepsin to 80 ml of the HCl solution.

10% pepsin stock solution: dissolve 100 mg in 1 ml of distilled water. Dispense into 90 μl aliquots and store at −20°C.

Protocol

1. Pre-warm the pepsin and 2 x SSC solutions to 37°C.

2. Place the slides in the 2 x SSC solution at 37°C for 10 min.

3. Remove from the 2 x SSC and place the slides in the pepsin solution at 37°C for 10 min.

4. Remove the pepsin and add PBS at room temperature for 3 min.

5. Remove the PBS and add the formaldehyde magnesium chloride solution at room temperature for 3 min.

6. Tip off the formaldehyde magnesium chloride solution and add PBS at room temperature for 3 min.

7. Dehydrate the slides by putting them through 70, 90 and 100% ethanol for 1 min each.

8. Tilt the slides upright and allow to dry.

Protocol 8.4

Standard protocol for *in situ* hybridization

Slides that have been prepared for FISH can sometimes be used immediately. However, hybridization may be more efficient if the slides are aged slightly beforehand. This can be done either by leaving the slides out on the bench overnight or by placing the slides in 2 x SSC at 37°C for 30 min. They may then be dehydrated by putting them through 70, 90 and 100% ethanol to standardize them (as outlined in *Protocol 8.3* above). The method assumes that the probes are already at a working dilution (note that some probes require dilution before use), and that they do not require a prehybridization step as is sometimes necessary for some chromosome paints. In addition, some probes are pre-denatured and therefore do not need to be denatured before hybridization.

Equipment

Coplin jars

Coverslips 18 × 18 mm

Coverslips 25 × 50 mm

Eppendorf tubes

Fluorescence microscope with appropriate filter sets, for example single-band-pass DAPI, Fluorescein isothiocyanate (FITC), double-band pass filter set, and so on

Forceps

Hotplate

Microscope slides with frosted ends

Pasteur pipettes, 1 ml

Surface thermometer

Waterbath

Solutions

20 x SSC

4,6-diamino-2-phenyl-indole (DAPI) 125 ng ml^{-1} or appropriate counterstain

Absolute ethanol, Analar-grade

Biotin–digoxigenin detection kit, if probe not directly labeled

Crushed ice

DNA probe, direct or indirectly labeled

Formamide (ultrapure quality)

Laboratory parafilm

Phosphate buffered saline (PBS)

Rubber cement (such as that found in bicycle repair kits)

Tween 20

Protocol
Preparation of samples and slides for FISH

The samples are processed and slides made in exactly the same way as for routine cytogenetics (see protocols in Chapter 2).

Denaturation of slides and hybridization

1. Add 5 μl of probe to the slide.

2. Cover with an 18 × 18 coverslip and seal with rubber cement.

3. Denature at 70–75°C on a hotplate for 2 min.

4. Allow to hybridize at 37°C overnight in a dark, sealed box containing damp paper to prevent the probe from drying out.

   Alternatively the probe and slide can be denatured separately as follows.

Denaturation of probe

1. If the probe is not predenatured, denature it by placing it in an Eppendorf tube and incubating it at 70°C for 5 min and then chilling on ice for about 5 min before applying to the denatured chromosomes.

Denaturation of slides

1. Make up the denaturation solution in a Coplin jar of 70% formamide and 2 × SSC adjusted to pH 7.0.

2. Heat the denaturation solution to 70°C in a waterbath.

3. Pre-warm the slides in an incubator at 37°C.

4. Add the slides to the denaturation solution in the waterbath and leave for 2 min.

5. Dehydrate the slides through ice-cold 70, 90 and 100% ethanol, leaving them for 1 min in each solution.

6. Tilt the slides upright and allow to air dry.

Hybridization

1. Add 5 μl of the probe to the slide.

2. Cover with an 18 × 18 coverslip and seal with rubber cement.

3. Allow the slide to hybridize at 37°C overnight in a dark, sealed box containing damp paper to prevent the probe from drying out.

Stringency washing

1. Fill a Coplin jar with SSC containing a drop of Tween 20. The concentration of SSC required depends on the stringency required for the probe used and ranges from 0.4 to 2.0 x SSC (see *Section 2.3*).

2. Just before washing, remove the rubber cement from all the slides and then remove the coverslips.

3. Place each slide in the preheated wash solution for 2–5 min (depending on stringency) at 72±2°C. (Note that the temperature of the wash solution should be directly measured using a thermometer; the temperature reading of the hotbath should not be relied upon.) No more than two slides should be washed at the same time since each slide will lower the temperature of the wash solution. The solution should then be allowed to reheat between washes.

4. Transfer the slides from the hot SSC to 2 x SSC Tween 20 at RT for 2–10 min.

5. Remove the slides from the SSC and stand them upright to drain. The slide should not be allowed to dry completely.

Detection of directly labeled probes

1. Add 10 μl of DAPI antifade or alternative counterstain and place a 25 x 50 mm coverslip on the slide.

2. Examine the slide using the fluorescence microscope.

Detection of indirectly labeled probes

The times of incubation and the solutions used here vary widely depending upon the detection kit. A very rough guide is outlined below.

1. Add about 100 μl of the fluorescently labeled antidigoxygenin and/or avidin to the slide, depending on the concentration.

2. Cover with a piece of parafilm and incubate at 37°C in the dark for 5–15 min.

3. Remove the parafilm.

4. Wash the slide through a series of three washes, either 2–4 x SSC Tween 20 or phosphate buffered solution, containing Tween 20, for 2 min each.

5. Add 10 μl of DAPI antifade or alternative counterstain and place a 25 x 50 mm coverslip on the slide then seal with rubber cement.

If the signal requires amplification:

6. Peel off rubber cement using forceps and then soak off the coverslip in 2 x SSC.

7. Remove the coverslip and soak in fresh 2 x SSC for 5 min.

8. Add about 100 μl of the fluorescently labeled antibody for the digoxygenin detection system, or the antiavidin for the biotin detection system.

9. Incubate at 37°C in the dark for 5–15 min.

10. Wash three times in the SSC/PBS Tween 20 solutions for 2 min each (the slides detected with digoxygenin can now be mounted with counterstain, as above).

11. Add about 100 μl of fluorescently labeled avidin for the biotin detection system.

12. Incubate at 37°C in the dark for 5–15 min.

13. Wash three times in the SSC/PBS Tween 20 solutions for 2 min each.

14. Mount with counterstain, as above.

Hints and tips

High background fluorescence: this is usually due to inadequate stringency washing, either the temperature of the wash solution was too low, or the concentration of the SSC solution was too high.

Poor signal: this may be due to the stringency of the wash being too great due to the temperature being too high or the SSC concentration being too low. It could also be due to insufficient denaturation of the slide or probe DNA.

Poor morphology of preparation: this is probably due to over denaturation of the target DNA before hybridization.

Appendix I

Glossary of terms

Acrocentric: Chromosome with the centromere near one end.

Acquired abnormality: Chromosome change which has arisen in abnormal cells during the course of disease.

Acute lymphoblastic leukemia (ALL): A progressive malignant disease characterized by large numbers of immature cells of the lymphoid series in bone marrow, peripheral blood and lymph node, spleen, and other organs.

Acute myeloblastic leukemia (AML): A malignant neoplasm of blood-forming tissues characterized by uncontrolled proliferation of immature cells of granulocytic lineage.

Acute non-lymphoblastic leukemia (ANNL): Any leukemia that is not of lymphoid origin.

Acute promyelocytic leukemia (APML): A malignancy of blood forming tissues, characterized by bleeding, scattered bruises, and proliferation of promyelocytes in bone marrow.

Adenocarcinoma: Any of a large group of malignant epithelial cell tumors of the glands.

Alleles: Alternative forms of a gene at the same locus.

Amniocentesis: Procedure for aspirating amniotic fluid.

Anemia: A decrease in hemoglobin concentration in blood below the normal range.

Aneuploid: A chromosome number which is not an exact multiple of the haploid set.

Aplasia: A lack of development of organ or tissue, or the cellular products of the organ or tissue.

Aplastic anemia: A deficiency of all formed elements of the blood representing a failure of the cell generating capacity of the bone marrow.

Auer rod: Abnormal, needle shaped, pink staining inclusion in the cytoplasm of myeloblasts and promyelocytes, which contain enzymes and represent abnormal derivatives of cytoplasmic granules.

Autosome: Any chromosome which is not a sex chromosome.

Basophil: A granulocyte characterized by a segmented nucleus and cytoplasmic granules, which stain purple when exposed to a basic dye.

B-cell (or B lymphocyte): A type of lymphocyte that expresses monoclonal immunoglobulin on its surface, and differentiates into plasma cells upon suitable antigenic stimulation.

Benign: A tumor or condition which is not malignant.

Bivalent: Structure formed between two homologous chromosomes via synapsis prior to the first meiotic division.

Blast cell: An immature cell which is a leukocyte precursor.

B lineage: B lymphocytes and their precursors.

Budd–Chiari syndrome: A disorder of hepatic circulation marked by venous obstruction, leading to liver enlargement and severe portal hypertension.

Burkitt's lymphoma: A malignant neoplasm common in African children, composed of mature B cells with distinct cytological features. The Epstein–Barr virus may be one of the etiological factors in the development of this lymphoma.

Carcinoma: A malignant epithelial neoplasm that tends to invade the surrounding tissues and metastasize around the body.

Centromere: A heterochromatic region in the chromosome which holds the chromatids together.

Chiasma: Formed during meiotic recombination by the crossing over of chromatid strands of homologous chromosomes.

Chimera: Individual whose cells arise from more than one zygote.

Chorionic villus sampling: Procedure used in prenatal diagnosis where chorionic villi are biopsied.

Chromatid: Strand of replicated DNA prior to separation during mitosis.

Chromatin: DNA complex and associated proteins which represent the state of genes in a nucleus.

Chromosomal aberration: An abnormality of chromosome structure or number visible under the light microscope.

Chronic granulocytic leukemia (CGL): A malignant neoplasm characterized by a proliferation of cells of granulocytic lineage (e.g. neutrophils, basophils and usually eosinophils) and also megakaryocytes.

Chronic lymphocytic leukemia (CLL): A neoplasm characterized by the proliferation of small, long-lived lymphocytes.

Chronic myeloid leukemia (CML): A malignant neoplasm, characterized by proliferation of cells of the granulocytic lineage with or without monocyte or megakaryocyte proliferation.

Clone: A cell line, which is derived by mitosis from a diploid cell, or propogated gene sequences from an identical parent gene.

Complementary: Two single strand nucleic acid molecules which form a succession of A/T and G/C base pairs in an antiparallel orientation.

Congenital: Present at birth.

Constitutional abnormality: A chromosome change which is present from birth.

Cosmid: A synthetic cloning vector for accommodating large portions of foreign DNA.

Cross-over: Occurs during meiosis, the exchange of genetic material between homologous chromosomes.

Deletion: Loss of a chromosomal segment.

Denaturation: The separation of complementary DNA by high temperature or alkali treatment.

Dicentric: Abnormal chromosome structure with two centromeres.

Differentiation: The process whereby a cell becomes committed to a certain lineage, also used to imply maturation.

Disseminated intravascular coagulation (DIC): A grave coagulopathy due to over-stimulation of the body's clotting and anti-clotting processes.

Diploid: Normal chromosome number in somatic cells.

Disomy, uniparental: Both homologs of a chromosome from the same parent are inherited, with loss of the corresponding homologs from the other parent.

Distal: Region distant from a particular area, such as a point furthest away from the centromere of a chromosome.

DNA: Deoxyribonucleic acid.

Dominant gene: Gene which carries a characteristic that over-rides recessive gene expression in a heterozygous individual.

Eosinophil: A granulocyte, larger than a neutrophil, with a bi-lobed nucleus and coarse cytoplasmic granules that stain intensely with eosin.

Eosinophilia: An increase in the number of eosinophils in the blood.

Epstein–Barr virus (EBV): The herpes virus that causes infectious mononucleosis.

Erythrocyte: Major cellular element of the blood (red blood cells), the cells that transports oxygen.

Erythrocytosis: An abnormal increase in the number of circulating red blood cells.

Erythroleukemia: A malignant disorder characterized by proliferation of erythropoietic elements in the bone marrow.

Erythropoiesis: The process of erythrocyte production.

Essential thrombocythemia (ET): Thrombocytosis consequent on neoplastic proliferation of megakaryocyte precursors.

Euchromatin: Nuclear DNA that remains relatively unfolded during most of the cell cycle allowing access to transcriptional machinery.

Gene: Linear collection of DNA sequences which are required to produce a functioning RNA molecule.

Fanconi's anemia: The commonest inherited form of aplastic anemia.

Follicular lymphoma: A nodular well differentiated lymphocytic malignant lymphoma, where neoplastic cells form follicles and distort the structure of the lymph node.

Genomic imprinting: Expression of parent specific genes or chromosomes in offspring.

Genotype: Genetic constitution of an organism.

Grade: An expression of the degree of malignancy of a tumor, high-grade being more aggressive than low grade.

Granulocyte: One of the leukocyte types, characterized by the presence of cytoplasmic granules, e.g. basophil, eosinophil, neutrophil.

Granulocytopenia: A decrease in the toal number of granulocytes.

Hematopoiesis: The production of all types of blood cells.

Haemostasis: The termination of bleeding by mechanical or chemical means or by the coagulation process of the body.

Hairy cell leukemia (HCL): A rare neoplasm, characterized by pancytopenia, large spleen and the presence in the blood and bone marrow of abnormal B lineage lymphocytes with many fine projections on their surface.

Haploid: Number of chromosomes present in normal gametes.

Hepatomegaly: Enlargement of the liver.

Hepatosplenomegaly: Enlargement of both the liver and spleen.

Heterochromatin: Regions of chromosomes, which remain tightly folded during the entire cell cycle, replicating late in S phase and not containing actively transcribed genes.

Heterozygote: Individual with a dominant allele and recessive allele on a pair of homologous chromosomes.

Histiocyte: A macrophage (large scavenger cell) which is stationary in the connective tissue.

Hodgkin's disease: A malignant disorder characterized by progressive enlargement of lymphoid tissue, splenomegaly, and the presence of Reed–Sternberg cells (atypical cells of uncertain lineage with multiple or hyperlobulated nuclei containing prominent nucleoli).

Homozygote: Individual with identical allelles (i.e. both dominant or both recessive) on a pair of homologous chromosomes.

Hybridization: Binding of nucleic acid sequences via complementary base pairing.

Ideogram: Diagram of the chromosome complement.

Idiopathic thrombocytopenic purpura (ITP): Bleeding into the skin and other organs owing to platelet deficiency.

Isochromosome: Chromosome abnormality where loss of one arm occurs (either p or q) and the remaining arm is duplicated (forming a mirror image of that arm).

Isodisomy, uniparental: Two copies of one homolog of a chromosome from the same parent are inherited, with loss of the corresponding homolog from the other parent.

Karyotype: Classified chromosome complement of a cell or individual.

Leukemoid reaction: Clinical syndrome resembling leukemia in which white cell count is raised in response to allergy, inflammation, infection or hemorrhage.

Leukocyte: White blood cell: five types classified according to cytological features: (a) mononuclear cells (lymphocytes and monocytes), (b) granulocytes (neutrophils, basophils and eosinophils).

Leukocytosis: Increase in the number of circulating white blood cells.

Leukopenia: Decrease in the number of circulating white blood cells.

Locus: Location of a gene on a chromosome.

Lymph node: Small oval structures in the lymphatic system, which filter lymph and mount an immune response to antigenic stimuli.

Lymphocyte: A leukocyte with predominantly agranular cytoplasm originating in the bone marrow or thymus, B cells and T cells.

Lymphocytopenia: Decreased number of circulating lymphocytes.

Lymphocytosis: Increased number of circulating lymphocytes.

Lymphoma: A neoplasm of lymphoid tissue.

Macrocytic anemia: Blood disorder characterized by impaired erythropoiesis and presence of large red blood cells.

Macrophage: Phagocytic cell of the reticulo–endothelial system derived from a peripheral blood monocyte.

Malignant: Tending to become worse, causing death. Describing a cancer that is metastatic; tending to recur even upon removal of the tumor tissue.

Maturation: The process by which a cell acquires characteristics of a more mature, or end stage in any given lineage.

Megabase: 1×10^6 base pairs of DNA.

Megakaryocyte: Bone marrow cell which produces platelets and releases them into the blood circulation.

Megaloblast: Abnormally large nucleated immature erythrocyte.

Megaloblastic anemia: Blood disorder characterized by production and proliferation of immature large and dysfunctional erythrocytes.

Meiosis: Cell division which results in reduction of chromosomes number during the formation of gametes.

Metastasis: The process by which tumor cells are spread to other parts of the body.

Microdeletion: Deletion of a chromosome, which is close to the resolution limits of the eye using a light microscope.

Mitosis: Somatic cell division.

Monoclonal antibody: An antibody produced by a single clone of cells, usually a hybridoma.

Monocyte: Large mononuclear leukocytes with an ovoid or kidney-shaped nucleus.

Monosomy: Loss of a whole chromosome.

Mosaic: Individual derived from a single zygote with cells of two or more differing genotypes.

Multiple myeloma: Plasma cell neoplasm which characteristically causes osteolytic bone lesions.

Mutation: Alteration in the genetic material.

Mycosis fungoides: Rare chronic lymphomatous skin malignancy resembling eczema or a cutaneous tumor.

Myeloblast: The earlier precursor of granulocytes.

Myelocyte: Immature white blood cell, which can be identified by belonging to the basophil, neutrophil or eosinophil lineage.

Myelodyplastic syndromes (MDS): A group of bone marrow diseases characterized by dysplastic (morphologically abnormal) and ineffective hemopoiesis.

Myeloproliferative disorders (MPD): A group of conditions characterized by endogenous proliferation of one or more hemopoietic components of the bone marrow.

Neoplasia: The process in which a genetically altered precursor cell gives rise to an abnormal clone of cells which show abnormality in proliferation and maturation.

Neutropenia: Decrease in the number of circulating neutrophils.

Neutrophil: A polymorphonuclear granular leukocyte, essential for phagocytosis and proteolysis.

Non-disjunction: Failure of a chromosome pair to divide during anaphase.

Normoblast: Nucleated precursor cell of the circulating erythrocyte.

Nucleotide: Purine or pyrimidine base attached to a sugar and phosphate group.

Oncogene: Gene sequence which can cause transformation.

Pancytopenia: Abnormal condition characterized by marked reduction in all major cellular elements of the blood.

Phenotype: Outward appearance of an individual.

Plasma cell: The most mature cell in the B lymphocyte lineage which secretes antibody.

Platelet: Smallest cells of the blood which are anuclear, essential for normal hemostasis.

Polycythemia: Abnormal increase in the number of erythrocytes in the blood.

Polycythemia rubra vera (PRV): Polycythemia consequent on neoplastic proliferation of red cell precursors.

Polymorphism: Genetic variation arising in the normal population.

Polyploid: Abnormal number of chromosomes, which is an exact multiple of the haploid number.

Probe: Labeled DNA or RNA fragment used to identify a complementary sequence in target cells using molecular hybridization.

Prognosis: The expected outcome of a disease.

Promyelocyte: Granulocyte precursor.

Purpura: Hemorrhage into the skin or mucous membranes.

Recessive: Gene which carries a characteristic that is only expressed in a homozygous individual for that gene.

Recombination: Process by which two homologous DNA duplex molecules exchange information during cross-over.

Reed–Sternberg cells: Neoplastic cell of uncertain lineage occurring in Hodgkin's disease.

Remission: A state in which all detectable evidence of neoplasm has disappeared.

Restriction enzyme: Enzyme which cleaves DNA at a sequence specific site.

RFLP: Restriction fragment length polymorphism: Recognition site for a restriction enzyme.

Reverse transcriptase: Enzyme which produces complementary DNA from messenger RNA.

RNA: Ribonucleic acid.

Satellite DNA: Repetitive DNA often found in heterochromatin.

Secondary constriction: Constriction on a chromosome other than the centromere.

Secondary leukemia: Leukemia attributable at least in part to prior radiotherapy, chemotherapy or exposure to other identifiable mutagens.

Segregation: Separation at meiosis of allelic genes.

Sideroblast: Abnormal erythroblast with a ring or iron containing granules around the nucleus.

Sideroblastic anemia: Anemia characterized by the presence of abnormal sidereoblasts in the bone marrow.

Sister chromatid exchange: Exchange of DNA within sister chromatids.

Southern blotting: Process used to filter transfer DNA fragments by size via gel electrophoresis.

Splenomegaly: Abnormal increase in spleen size.

Syndrome: Combination of features which is non-random.

T cell (or T lymphocyte): A lymphocyte involved in cell-mediated immunity, the maturation of which is influenced in the thymus.

Teratogen: Agent which causes genetic malformations.

Tetrasomy: Two extra copies of a chromosome in a cell.

Thrombocytopenia: Reduction in the number of circulating platelets.

Thrombocytic purpura: A bleeding disorder characterized by a decrease in the number of platelets resulting in multiple bruising (also called idiopathic thrombocytopenic purpura: ITP).

Thrombocytosis: Increase in the number of circulating platelets.

T-lineage: T cells and their precursors.

Transcription: Messenger RNA production from the DNA template.

Translation: Process where protein is synthesized from a messenger RNA sequence.

Translocation: Transfer of chromosomal segments between chromosomes.

Triploid: Cell with three times the haploid chromosome number, i.e. 69.

Trisomy: Extra copy of a chromosome in a cell.

Vector: Plasmid, phage or cosmid into which foreign DNA can be inserted for cloning.

Tumor: A neoplasm growing as a discrete mass of neoplastic tissue.

YAC: Yeast artificial chromosome used for the cloning of large DNA segments.

Zygote: Fertilized ovum.

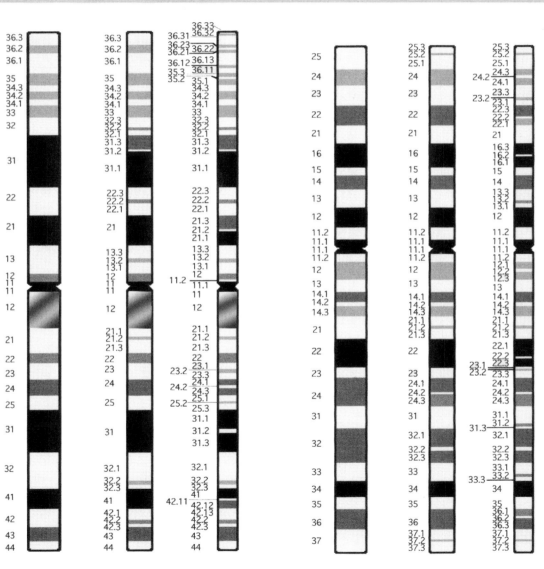

CHROMOSOME 1   CHROMOSOME 2

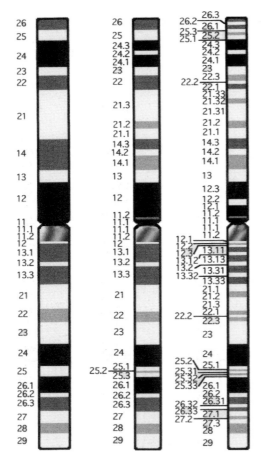

CHROMOSOME 3

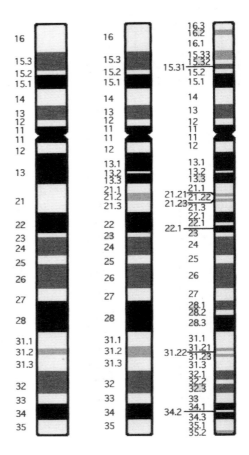

CHROMOSOME 4

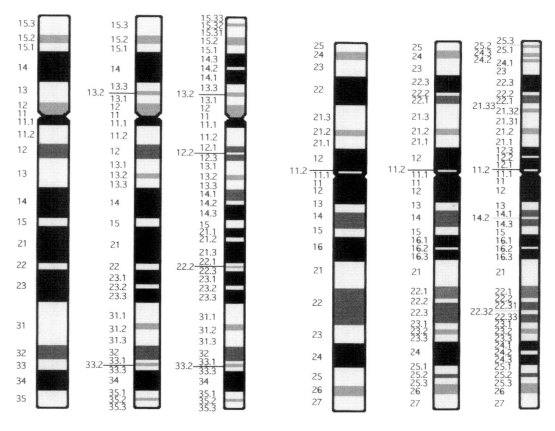

CHROMOSOME 5   CHROMOSOME 6

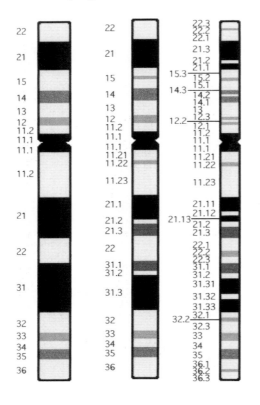

CHROMOSOME 7

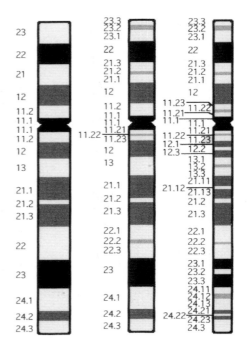

CHROMOSOME 8

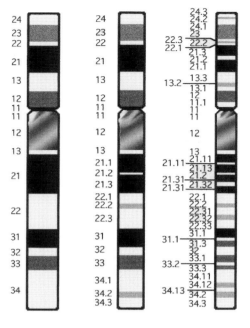

CHROMOSOME 9

CHROMOSOME 10

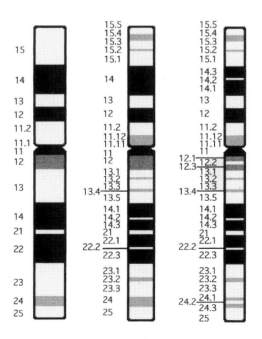

CHROMOSOME 11

CHROMOSOME 12

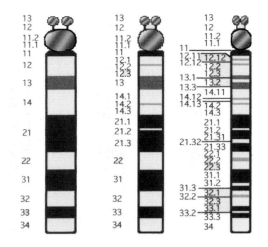

CHROMOSOME 13

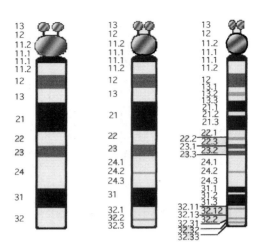

CHROMOSOME 14

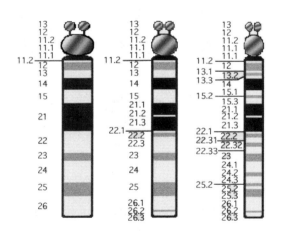

CHROMOSOME 15

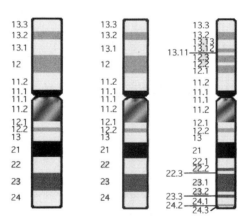

CHROMOSOME 16

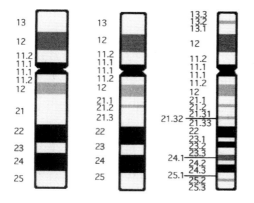

CHROMOSOME 17

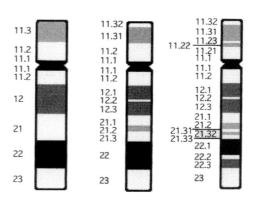

CHROMOSOME 18

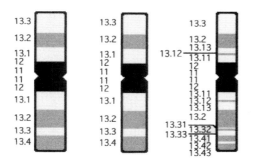

CHROMOSOME 19

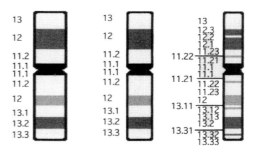

CHROMOSOME 20

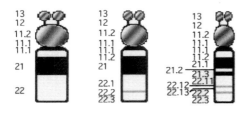

CHROMOSOME 21

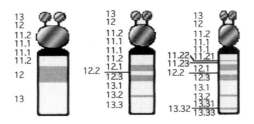

CHROMOSOME 22

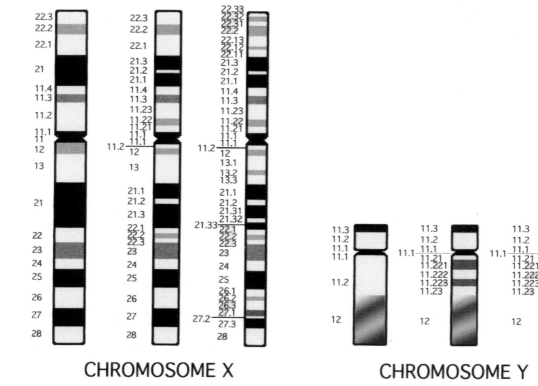

CHROMOSOME X

CHROMOSOME Y

Figure legend for Appendix II

Ideogram based on the ISCN (1995) demonstrating the G-banding patterns for normal human chromosomes at different banding resolutions. The left-hand chromosome represents a chromosome at a 400-band level, the center chromosome shows a 550-band level, and the right-hand chromosome in each set represents an 850-band level. The ideogram illustrates the varying intensities of the G-bands, for instance, note the light G-bands on chromosome 19p and 19q as opposed to the darkly staining centromere in this chromosome. Note also the angled striped areas near the centromeres of chromosomes 1, 3, 9, 16, and q arm of Y, and the short arms of the acrocentrics, chromosomes 13, 14, 15, 21 and 22, these represent areas of the chromosome which can be variable in the normal human karyotype.

Index